GLOBAL ENVIRONMENTAL ISSUES

Encyclopaedia of Environmental Pollution and Awareness in 21st Century Series

GLOBAL ENVIRONMENTAL ISSUES

Edited by

V.K. Prabhakar

ANMOL PUBLICATIONS PVT. LTD.

NEW DELHI - 110 002 (INDIA)

ANMOL PUBLICATIONS PVT. LTD.
4374/4B, Ansari Road, Daryaganj
New Delhi - 110 002

Global Environmental Issues

First Edition, 2001

ISBN 81-261-0906-8

PRINTED IN INDIA

Published by J.L. Kumar for Anmol Publications Pvt. Ltd., New Delhi - 110 002 and Printed at Mehra Offset Press, Delhi.

CONTENTS

PREFACE

Today, the cry of 'pollution' is heard from all the nooks and corners of the world, and pollution has become a major threat to the very existence of mankind on this earth. It is the major challenge of our times.

The pollution of various resources has gone to such an extent that we are unable to breathe fresh air and drink fresh water. On one hand the advancements of science and technology have added to the human comforts by giving us automobiles, electrical appliances, supersonic jets, space crafts, better medicines, better chemicals to control harmful insects and other pests, etc., but, on the other hand, they have given us a very serious problem to face that is pollution. Pollution can be defined as an undesirable change in the physical, chemical, or biological characteristics of the air, water, or land that can harmfully affect health, survival, or activities of humans or other living organisms.

Pollution does not have to cause physical harm, but it may merely interfere with human activities. Chemicals spewed into the air or water from an industrial plant may be harmful to humans and other organisms living near the plant. If expensive pollution controls are required, the plant may be forced to shut down.

India faces serious problems of environmental pollution. Water pollution from untreated community and industrial wastes flowing into our rivers and streams, increasing use of fertilizers and pesticides, dumping of organic wastes etc., are issues that need to be tackled on the priority. Air pollution, long regarded as the bane of industrialised nations looms large over our country.

Here, we are presenting this *Encyclopaedia of Environmental Pollution and Awareness in Twenty-first Century* in fifty volumes with a view to make people conscious about the seriousness of pollution in the present time and its remedies.

Editor

1 Introduction

A world strategy for environmental protection would likely favour from an unanimous view of what needs to be done. But is there a universal view on the major global environmental threats, as a basis for taking wise decisions? Based on individual knowledge in the field of environmental protection, and visions of how a balanced ecosystem might look, there is likely to be quite a number of deviating opinions in different countries. To a considerable extent, these opinions are probably influenced by public institutional and political viewpoints. Which are the world's "top three" environmental issues?

An idea of the diversity surrounding this concept was developed at the ECOLOGY 89 conference in Gothenburg in August 1989 by surveying representatives from several of the international organizations. Their responses are summarized below.

United Nations Economic Commission for Europe (ECE)

1. Coniferous forest decline.
2. Mismanagement of hazardous wastes.
3. Transboundary pollution of air and water;

European Environment Bureau (EEB)

1. Decreasing biodiversity,
2. Climate changes,

3. Population growth;

European Trade Union Confederation (ETUC)

1. Deterioration in the work environment,
2. Gradually-lowered living standards,
3. Unemployment;

United National Food and Agriculture Organisation (FAO)

1. Uncontrolled increase in world population,
2. Misuse of renewable nature resources, leading to depletion,
3. Degradation of the terrestrial and biotic environment;

Greenpeace

1. Global toxic pollution,
2. Environmental and biological degradation,
3. Human ignorance and fear of change;

International Labour Organisation (ILO)

1. Lack of preparedness for major hazards,
2. Risks in the work environment,
3. Conflicts between the North (looking for strategies to solve environmental problems) and the South (requiring strategies to solve their development and poverty problems, while also achieving their environmental objectives):

Organization for Economic Co-operation and Development (OEC D)

1. Climate change,
2. Toxic substances and wastes,
3. Economic-political destabilization in developing countries;

United Nations Environment Programme (UNEP):

1. Risks to human beings,

2. Risks to living resources,
3. Risks to natural resources;

World Health Orgnization (WHO)

1. Impacts of development policies on health and the environment of development policies,
2. Pathogenic factcrs in the environment,
3. Psycho-social factors in the environment effecting quality of life;

World Meteorological Organization (WMO):

1. Climate variability and change,
2. Natural disasters,
3. Increased atmospheric pollution;

World Bank

1. Destabilization of ecological balances,
2. Depletion and destruction of biological resources,
3. Production and release of persistent-hazardous compounds;

Regardless of national or organizational origin, it should be possible for every-one to agree, in general terms, on what type of environmental problems should be considered more serious than others. One point of departure is to find a distinction based on:

— threats to plant, animal and human life in various time and geographic perspectives;

— the origin and hazardous character of pollutants; and,

— the potential of reversing the problem.

Environmental problems imposing threats ,to life on Earth, leading to extinction of whole species, ought to be considered major threats. The spread of man-made pollutants, that are bioaccumuable, persistent and toxic, are often the cause of such threats, since these pollutants are often hazardous at low concentrations. Pollutants of natural origin usually lead to environmental problems, if present in excessive concentrations. However, problems of these kinds are often reversible, if it is possible to initiate control of emissions of these elements prior to the pollution becoming too grave. Based on these distinctions itmight be relevantto make the following statements on suitable criteria;

— non-reversibleenvironmental problems are more serious than reversible ones-the reversibility criterion;
— long-term promblems are more serious thanshort-terms promblems-the time criterion;
— problems with a large geographical extent are more serious than those which affect only smaller areas-the area criterion; and,
— problems on which only partial or incomplete knowledge is avaiable should be given more attention than those that are better understood-the precautionary criterion.

Based on these elementary principles, the following prioritization of ten well-known environmental problems might be envisioned:

— loss of valuable plant and animal species;
— depletion of the stratospheric ozone layer;
— global warming;
— pollution by hazardous chemicals, processes and wastes;
— acidification of soils and water;
— deforestation;
— desertification;
— eutrophication of waters; and,
— degradation of land resources.

The differing views of the various international organizations described above regarding major threats to our environment posed by different human activities reflect a political decision-making dilemma. what are the main environmental problems to attack, and that types of measures should be carried out initially? These questions cannot be avoided when deciding about envirronmental conttrol programmes. It is important to stress that there is not always a contradication, as earlier believed, between economic growth and satisfactory environmental quality.

However, it is necessary to further elaborate on the definition of economic growth. Economic development should be considered a positive force, and a necessary condition for environmental protection and vice versa. Accordingly, the nature of the policy response required to safeguard the environment and to assure the sustainability of economic growth, must be considered.

Not all environmental problems are equally serious. Thus, in some cases, measures for protecting the environment do not need to be as far-teaching as for other problems. There may be areas in which environmental improvement are needed more-even if the improvement may not be as great. It is commonly agreed that available financial resources must be allocated to those areas where the greatest environmental improvement can be achieved.

The only way to achieve this goal is through a process of setting priorities. It is very important that decisions taken on future environmental-pollution abatement measures are based on a balance of what is technically feasible, economically reasonable, and ecologically motivated, and that the environmental impacts are assessed with regard to the origin of pollution. However, environmental efficiency must also play a more dominant role in environmental decision-making in the future. This is because higher priorities must be given to those areas where the environmental risks and associated time constraints are considered to be the greatest. The complexity of ecological responses and their relationships to environmental stresses aggravates the selection of suitable remedial measures. However, prior to any decision-making process being carried out, some basic assumptions often forgotten need to be satisfied. An example is setting desired environmental goals, and the driving forces to attain in the from of various incentives need to be articulated.

Environmental goals must be clearly defined, and a distinction must be made between emission goals. It is particularly important to stress the need for elucidating the desirable state of the environment, based on different user aspects. Without focusing on specific environmental conditions, future assessment of whether or not our pollution abatement efforts were sufficient cannot be made. Recently, it is become obvious that other incentives, apart from legislation, are likely to be a driving force in future environmental work, including:

— the need to further improve the work environment;
— the need to satisfy growing environmental awareness among consumers, and the greater consumer demands for environmentally-friendly products;
— the need to develop a better "environmental image" for companies and authorities, in order to assure the recruiting of competent and motivated persons; and,
— the need to strengthen international cooperation and agreements for further environmental efforts.

Future environmental decisions will likely become more difficult mainly because of the complex interrelationships between causes and effects of many environmental problems. Thus a systematic approach for setting priorities will be a useful tool in environmental decision-making, and for guiding industrialists and politicians at the highest levels. By a continual testing and evaluation of environmental protection efforts the need fir a general framework for such an approach might be envisioned. A rational, careful handling of environmental issues, based on a common strategy for preventive measures, will be a strength in all parties' environmental policies. Hopefully, it will raise the credibility of all the active participants in facing the challenge of safeguarding the environment.

The concept of sustainable development generally has met with broad international acceptance. It implies the agreement with necessary changes in people's lives and business philosophies, based on economic and technological development, to improve the environmental situation.

A balance must exist between technological development, prosperity, and the conservation and improvement of the environment. New incentives must trigger the development of social structures in all countries. It should be stressed that a sound economy will be likely to accelerate the progress of environmental protection efforts in any country.

Therefore, the management of sustainable development rests upon the participation; of all people in very nation, and especially on concerted efforts carried out by the world's industries.

Environmental problems are not caused solely by the consequences of humanity striving for a higher standard of living and material wealth. They are also dependent on organizational and institutional failures to insert effective counter-measures, but also upon human agreed. Thus, to enable inclusion of a sustainable development in society's philosophical base, a multidisciplinary approach is needed in a close interplay between various actors, environmental issues and remedial measures.

Ever since the expression sustainable development was first introduced in the report from the World Commission of Environment and Development it had been subject to almost endless debate and discussion about its meaning and scope. The debate still continues.

Whatever the conclusions of such discussion, it is important to elucidate suitable indicators of sustainable development, thereby making it possible to assess whether future human development is in compliance with sustainability or not. It is probably not preferably to use only a single-criterion approach.

Rather, there should probably be several indicators-anticipating the predicted trends, but not in too much details. Further, the indicators should cover social, market and ecological concerns, and preferably describe *inter alia*:

— standard of living;
— 'green' GNP;
— material flows; and,
— biodiversity.

Different points of departure can be used for the surveillance of indicators of sustainable development. Examples, with regard to ecological aspects, could be:

— changes within natural; patterns of variation;
— production equalling consumption; and,
— maintained biodiversity.

Throughout history, humanity has altered its surroundings for its own benefits. Therefore, it seems not possible to establish steady-state conditions that will prevail forever. The evolution must of on, but bearing in mind also the maintenance of ecological balances insofar as possible. Thus, it might be necessary to temporarily accept the continuation of a negative trend in some aspects. It should also be stressed that it is not realistic to foresee a future with no residual wastes.

Few previous cultures or ecosystems have succeeded in totally circulating energy and material flows. However, humanity has a fair chance to accomplish this goal to a much greater extent than at present. In order to ensure sustainable development, it is necessary to pay attention to economic efficiency and equity, both within and between generations.

There are no formal proofs. However, it is possible that both objectives can be achieved by ensuring that the next generation has access to the same ecological productive potential as our generation. This requirement can be formalized as follows: this generation should ensure that it passes on to the next generation a stock of assets no less than the stock it has inherited. The assets in question comprise all forms of capital. It should be recognized that environmental capital can just as much be considered capital in this sense as man-made capital.

Environmental capital includes not just stocks of oil, coal, gas and minerals. Rather, it also includes, e.g., the ozone layer, the protective functions of forests and wetlands, the self-purification functions of rivers and oceans and the store of biological diversity. Ensuring sustainable development is to 'maintain the capital stock' or, in other words, to 'maintain a full repairing lease'.

As long as the total of capital stocks is maintained, the stock of environmental assets may be able to be reduced in the short term, in favour of building up the stock of machines and infrastructure, and human skills and knowledge, to better be able to cope with future environmental problems, to increase the environmental assets in the long term. The notion of trading off environmental and other capital is not readily accepted by conservationists. The Brundtland Commission advanced the idea that it is not necessary to trade off the environment and development. This perspective might be too simplistic. It may be necessary to face trade-offs, at least during the foreseeable future.

However, there are major qualifications to the idea that environmental capital can be decreased as long as man-made capital and human skills are increased. It is important to consider:

- environmental assets frequently are being exploited without compensation, through increases in the amounts of other forms of capital-away of 'living off our capital';
- some forms of environmental assets are not readily substituted with other forms of capital. Examples include the store of biodiversity and the ozone layer. For these types of resources, there must be an 'environmentally-safe minimum standard'—a quantity and quality below which further exploitation should not be allowed; and,
- when 'trading-off' the various types of capital, without managing environment and development - it is vitally important to value the environmental assets correctly.

Subject to these very important considerations, sustainable development does not mean zero economic growth. It means growth within bounds set by the need to maintain the structure and function of our environmental capital. It also means a change in the manner in which growth is maintained, based on these prerequisites. In this context, sustainable development should be regarded as a continuous process, advanced by a stepwise approach; it is not an endpoint in itself.

It is wrong to consider sustainable development chiefly a matter confined to developing countries. In fact, the way the developed countries currently produce and consume resources is causing many global environmental problems.

It may be paradoxical for people in modern industrial economies to complain about environmental degradation taking place in other countries in the early stages of their development, when they do not themselves first demonstrate a capacity to become self-sustaining in their own patterns of production and consumption.

Even if they can achieve some reduction in the intensity of energy and material uses, developed countries may still remain the main contributors to environmental degradation. Further, if the best management practices currently followed in the North were universally adopted, the world would probably experience an increased environmental crisis.

Developed countries must put their own houses in order if they intend to preach to developing countries about the sacrifices they must make for out common future. Today's advanced economies do not have all the answers. A massive interdisciplinary research effort must be undertaken, bearing in mind that there will not be one single solution.

Given the extreme variability of local problems and solutions, a plurality of approaches will serve us best. Aid policies in developed countries have not always encouraged the promotion of environmentally-begin practices, based on local conditions and experiences. Industrial and other development projects in developing countries have often been based on conventional, and sometimes even obsolete approaches characterized by high environmental impact and consumption of resources.

Emerging technologies have the capacity to generate new, ecologically-sound patterns of development for all countries. It will be the emerging technology protagonists of the ongoing ecological revolution that likely will enable a shift to lower energy and raw material usage in an economics.

With the increased ambition to help developing countries to improve their environmental conditions, developed countries should not forget that developing countries, in their need to get access to scientific and technological advances, want to retain their freedom to choose their own developmental path.

But such development will not be possible with less than early and active participation of villagers and local authorities in the developing countries, based on a respect for their cultural heritage. This will favour a credible dialogue with developing countries, so that their rich and diversified culture can make its own contribution to the global response.

Although much progress has been made in municipalities to establish environmental control systems, much remains to be done. Legislation alone is not sufficient. Municipalities need to be more active in the entire filed of environmental protection-planning, implementation and control. The practical problems associated with designing and enforcing measures must be overcome.

Clear and understandable data must be developed on the environmental protection performance of existing and upcoming remedial measures. The critical issues of traffic management and waste handling on a municipal level may be considered the major challenges to be faced. The concept of waste handling is further shortly elaborate here, as an illustrative example and show-case of sustainable development of the necessary interplay between actions that need to be taken by citizens, authorities and private companies. A means must be found to enhance the ability of municipal departments to implement, finance and control recycling, treatment and disposal facilities, and to manage public acceptance of these facilities. A municipal strategy in terms of sustainable waste handling will affect almost all sectors of society, the most important being:

— *Production*—the challenge is to introduce "cleaner technologies and behaviour" that produce fewer waste discharges, and generate products that do not cause environmental pollution in the later stages of their life cycles-many of the raw materials, chemicals and manufacturing processes used today will have to be replaced by environmentally-friendlier alternatives;

— *Consumption*—the challenge is to convince people to buy products produced with an environmentally-friendly technology that does not harm the environment during their life-cycle, and that can be recovered or treated without polluting the environment when they are discarded; and,

— waste handling- the challenge is to:

— encourage waste avoidance by means of recycling and pre-separation at source,

— Promote recovery of material in municipal refuse,

— increase energy recovery from waste, in terms of better power generation, cogeneration and new development of heat demand around refuse treatment facilities,

- — cost-effective remediation of abandoned waste-disposal sites,
- — education for a broad target audience, including teachers in primary and secondary schools, municipal employees, politicians and decisions-makers, and,
- — communication for the broadening of public participation as an invaluable part of a planning and implementation programme.

It is vital concern to early identify the obstacles in resource recovery routes and waste handling. Combined efforts between municipalities, industries and selected waste companies might help to make an entire system of waste handling operational before its full introduction. Experiences from such systems might be used as guides for source separation and waste handling in neighboring communities, and favour establishment of inter-municipal waste- handling systems.

Effective municipal waste-handling system must rest on public participation, and a deep understanding of their responsibility and on the respect of citizens. Municipalities must identify early their citizen's roles sin the chain of events, from production to waste handling. From the viewpoint of citizens, the municipality have an *initial responsibility*, in terms of planning, enforcement and controlling production, a *second responsibility* in regard to information about product use and separation of household waste, and a *third responsibility* to guarantee proper caretaking and safe waste-disposal.

Combatting environmental pollution is specifically a business challenge. Increasing emphasis must be given to solve remaining environmental problems. Business ambitions in the field of environmental protection should be based on a redefinition of wealth, and the way that the business sector can contribute to economic growth. Environmental protection must be viewed as an investment for the future, with due consideration of competition and efficiency.

New incentives for these elevated environmental ambitions have emerged during the last few years- one being growing consumer demand for environmentally-friendly products, another being growing worker demand a high standard in the work environment, and yet another being upcoming legislation on product-related issues, e.g., producer liability.

These are important prerequisites to gain a good "environmental image" for attracting and recruiting skilled, environmentally-committed employees. Decision on complementary measures can only be made by the companies themselves, because of their own special viewpoint and knowledge about specific on-site conditions. Therefore business may take a lead on development of an environmentally-sound society, based on the *Precautionary Principle* as a base for environmentally-sound product development.

The business environment concept has to be broadened; a new philosophy must be introduced to advance from the old *"end of the pipe"*-approach to a new concept *"through the entire product cycle"*. The new concept, sometimes referred to as product stewardship, should incorporate all of the necessary steps, considering environmental impacts from the extraction of raw materials, through manufacturing process itself, use of the products, recovery and refuse of wastes to their ultimate disposal- the so-called "5R" strategy.

Health and safety aspects, in terms of a better work environment, must also be considered as well as material efficiency and its inherent environmental hazards and product performance. These are all components to form an *integrated pollution prevention and control* strategy. A key tool in these achievements is a high degree of coordinated assessment of *Best-Available-Technologies* to meet emission standards as well as of environmental quality objectives for agreed dangerous substances.

It is essential that such techniques be proven and available, and shown to be industrially viable in terms of technology as well as economics. In order not to inhibit technological department, techniques other than those identified should be equally acceptable provided their use achieve equal or better environmental performance.

An important step in the efforts to optimize the manufacturing processes is by managing *material balances*. Cost/benefit analyses are also a useful tool in environmental decision-making for guiding industrial leaders, even though such an approach had encountered some practical difficulties in validating the different outcomes in ecological terms.

Industry ambitions to become environmentally-friendly in various aspects will mainly depend on the development of *clean and low-waste technologies* and instruments for product policy as an important coornerstone. The application of this endeavour is also motivated for several economic reasons.

In a number of cases, it has been shown the introduction of clean technologies is economically profitable, and that investments in such technologies is economically profitable, and that investments in such technologies generally have a positive impact on the overall technical development and employment in a firm.

Companies, that have the ambition to be successful in the future, will have to take independent and responsible actions, based on their know-how and technology, to achieve environmental protection within reasonable economic constraints. Improper handling of hazardous wastes might be a dead-end for future business developments trying to move in an environmentally-friendly direction.

Therefore considerable efforts must be placed on introducing proper waste minimization programmes. Methods must be found to create a long-term demand for recovered raw materials based on steering recovery processes towards achieving the optimum mix of environmental efficiency, cost efficiency and product quality. Environmental protection, in terms of waste handling, is in a broad sense "management of scarce resources".

The role of trade and retail companies is specially important in completing the circle of efficient waste handling, in terms of pricing. It ought to be based on the full-cost accounting principle, a packaging and storage philosophy that uses a minimum of unnecessary and non-recyclable materials, as well as relevant consumer information systems on the environmental characteristics of their products.

Adopting a "cradle-to-grave" approach poses a major challenge to the business sector. However, business is not always in full control of all phases of their product's environmental impacts during their life cycles, because of, e.g., restrictions and bans of using specific raw materials and chemicals, and charges on pollutant emissions and wastes. Hence, business needs to identify its specific role and responsibilities in this context as well as influencing other parties of concern in the various stages of this cycle.

Future environmental work to be undertaken by various enterprises needs an integrated approach which engages all of the employees in a company. It is to be emphasized that company leaders and managers have a special responsibility to strengthen environmental commitments in accordance with the company's environmental policy, and promoting the importance of considering environmental matters among themselves and the workers. The environmental work ought to be an integral part of all company activities. A business approach to environmental management can include the following elements:

— stimulating Company Boards to routinely deal with environmental issues; environmental care should be institutionalized within enterprises as effective administrative measures to promote environmentally-oriented technology and products-the status of environmental work must be enhanced, based on responsibility and openness;

— including environmental issues in the planning of new processes and projects;

— maintaining an updated plan for environmental protection within the company;

— supporting, training and recruiting relevant competencies within the staff handling environmental matters, or to consult outside expert-this is of vital importance in achieving the demands in terms of the permits issued by the authorities;

— taking precautionary actions, and striving for further improvements in the processes and good housekeeping of treatment facilities;

— critically assesing permits, control programmes, and various forms of regulations affecting a company, in order to elucidate the existence of uncertainties that should be corrected to successfully carry out environmental protection;

- — inventorying and minimizing the risk of environmental accidents, occurring in both transportation and storage of goods, as well as maximizing the protection and safety of buildings; and,
- — reconsidering an expanded environmental insurance system concerning environmental protection liability.

Another component of future business activities is the ability to fulfil the ambition of practising preventive environmental measures, according to the so-called Precautionary Principle. However, such an ambition must rest on the inclusion of many new components in the industrial decision-making process.

Therefore, a new category of environmental strategists might be needed in companies. In large multinational companies, such professionals may even be a part of the permanent staff. These individuals may also be permanent members of the Board of Directors. Environmental strategists should preferably cover the following elements:

- — ancipating environmental laws and regulations;
- — forseeing the possible technological progress with bearing on the environment;
- — evaluating the business consequences, with regard to public and political pressures, for environmentally-sound product development;
- — analyzing "the market", in terms of green consumer demands, and threshold values in their willingness to pay for environmentally-friendly products; and,
- — guiding the advertising of a company's environmental ambitions and products in a manner consistent with marketing ethics.

Guiding principles for an environmental strategist working in the business sector should be realization of the following four cornerstones, developed from experiences of existing business environmental work:

— from *local to global* problems, with regard to changes in the spread of environmental pollution;
— from *simplicity to complexity*, with regard to changes in the features of environmental problems;
— from *dilution to recovery,* with regard to changes in the corporate pollution abatement efforts; and,
— from *general imagination* of environmental threats to *environmental impact calculations,* with regard to the needed changes in accuracy and reliability in environmentally-sound product development.

Environmental strategists should also be able to identify necessary steps in a planning procedure about available and additional needs relating to:

— know-where;
— know-when;
— know-how; and,
— know-who

In regard to guiding members of the Board in how to proceed to fulfil the environmental ambitions. An important tool for environmental strategists is the ongoing use of environmental audits. However, in using this tool an additional dimension ought to be included-the ecological aspects.

Ecological considerations must be an integral part of the pre-audit activities, especially during the phases of selecting criteria and assigning priorities. Ecological aspects must also be considered in connection with post-audit activities and during follow-up activities. Examples of contributions of ecological origin to environmental audits could be to:

— assess load-response relationships and ecological constraints imposed by pointsource pollution to air, soil and water;
— evaluate major threats to the environment from a company's overall operations;
— elucidate whether restrictions on production and permissions given by authorities are valid, based on ecological perspectives;
— set the framework and practical procedure for quality assurance in the purchases of raw material and choices of construction concepts of environmentally-friendly products;
— clarify the company's view on environmental impact in different time perspectives and scales, as a consequence of their own operation; and,
— formulate the company,s environmental policy.

It should also be possible for an environmental strategist to add new aspects to what a company can achieve with a given investment devoted to fulfilling authority demands. The goal is instead to find alternatives, indicative of more environmental-and cost-effective ways of conducting environmental protection work.

There is on one isolated national solution to many environmental problems facing the go\lobe. Nevertheless, every country basically must develop its own strategy on how to solve their environmental problems, based on their own superior knowledge of local conditions.

However, an environmental policy must be in accord with the fluctuations in public opinion and political will. In many cases, a lack of effective environmental programmes is not due to economic constraints. Rather, it depends largely upon difficulties in setting the right priorities in due time.

A national approach for solving environmental problems must be formed from comprehensive planning and based on preventive environmental efforts. Such an approach much gain the full support and commitment of all levels of society, and may include the following components:

— a national environmental management structure, ensuring full integration of ecological aspects into all social sectors;

— environmentally-sound legislation, with ecosystem specific land use planning and conceptualization of sustainable energy and economic futures; and,

— employment guarantees, in which the generation of employment is linked as much as possible to ecological regeneration.

In parallel with business ambitions to introduce environmental audits as a tool in preventing further environmental pollution a community environmental audit may also be carried out by responsible governmental authorities.

Major motives for this are the growing importance of nonpoint-source pollution as a consequence of consumer behaviour and buying patterns, which are an increased burden on governments to take a more active role in solving future environmental problems.

It is not unexpected that these demands likely will be raised in the future—every taxpayer certainly wants to know how authorities use their tax resources for environmental work, and the results of their achievements. A community environmental audit should highlight those issues not normally addressed in routine authority performances, and that specifically reflect public-related concerns. Examples are:

— information about the state of the environmental, and the proportion of polluting inputs from various sources;

— description of present and planned environmental protection programmes, and expectations of environmental improvements and summary of current achievements; and,

— information about product eco-labelling systems in use.

A growing environmental awareness calls for increased efforts by authorities, in terms of information activities. It is often frustrating to the public to have the feeling that authorities are not openly giving access to general environmental information.

Suitable and well-designed monitoring programmes based on review of previous data are key activities that generates information of public interest. Monitoring activities must not be used as an alibi for doing nothing. They should be continously evaluated and updated. Parameters reflecting important present and future environmental trends should be included in the monitoring programmes as early as possible.

It is preferable to conduct monitoring activities as an integrated air, soil and water-based approach. In order to gain insight into the concept of target and critical loads with their inherent natural variations, monitoring programmes must include some basic parameters that will be analyzed routinely over the long term.

It is also important to increase the predictive power of monitoring information by reducing unnecessary uncertainties, and then formulating the presentation of the results in a way that take full account to the level of public understanding.

Protecting the global environment calls for a determined and concerted international response, and for early world-wide adoption of suitable remedial strategies. This is both a formidable task and a challenging undertaking. Joint international actions have to be taken—initially among the Western developed countries—to gather knowledge on clean technologies and environmental techniques for meeting the goals of environmental compatibility.

Ways and means have to be found to transfer technologies and resources to developing countries, which add to their ongoing environmental programmes. Special attention should be directed towards East European and developing countries, to assist them to move towards a sustainable, environmentally-sound development and clean-up of already-degraded environments.

The urgent need to safeguard our environment for future generations should link together people from different societies and cultures to establish a global master plan based on, not only as usually environmental aspects, but also socioeconomic development.

These international efforts must also integrate every country's sovereignity in a climate of mutual understanding. especially, the interests and needs of developing countries to sustain the growth of their economic, financial and technological requirements for meeting environmental needs ought to be guaranteed, in order to retain their freedom. Several features should be part of such a master plan. Some of the most important features are listed below, namely on:

— focusing initially on common global concerns such as environmental problems in the atmosphere, the tropical forests, the deep sea and the polar regions;

— generating confident global management, based on ecologically-sound global trade and consumption patterns;

— assisting the universal quest for the best-available-technology by encouraging effective technology transfer and cooperation, and promoting appropriate local testing and adaptation;

— understanding the immediate and real needs of developing countries by supporting a local process towards environmental progress, promoting low-cost resource conservation, stimulating environmentally-sound urbaniza-tion and regeneration of village ecosystems including the strengthening of village democracy and self-reliance; and,

— enhancing environmental education, with regard for differences in people's backgrounds, knowledge and cultural; traditions.

Joint international efforts to ease the burden on environmental resources must be based on sustainable economic development taking into account effective environmental protection, careful and appropriate resource management and family planning. But it is also equally a matter of assuring necessary economic growth in those countries lagging behind as of ensuring the necessary educational level.

Economic constraints could be overcome to some extent by an early re-orientation of the socio-economic planning process towards environmental realities in the setting of societal objectives. However, an important prerequisite for this is the opportunity for full access by any country to vital environmental information, presented in various relevant native languages.

Addressing environmental problems on a global scale necessitates an acceptable solution to the international debt crisis. Means of funding international environmental-aid programmes must be sought as well as generating enhanced consideration of the concept of debt-for-nature swaps as a means of satisfying earlier debts.

Financial institutions such as the World Bank, the International Monetary Fund and the International Finance Corporation should be more deeply involved in assessing the environmental impacts of their funded project at an early stage. It is also of vital importance to improve the investment conditions in those countries demanding further growth of their various business sectors, with associated transfers to technology.

Such an opening of new markets for found environmental protection programmes is advanced by assuring neutrality in the business sector with regard to competition. Therefore, it is necessary to seek harmonization of basic environmental legislation not to dislocate activities harmful to the environment in countries with less-stringent regulations, which result in new environmental threats. The gradual deterioration of the environment lends compelling weight to the need to find a balance between human development and exploitation of natural resources.

Some people still claim that the environmental alerts are somewhat exaggerated. Forgetting about the many frightening but illustrative examples of some of their consequences, namely that;

— only half of the world's population breathes healthy air;
— less than half of the world's rural population has clean water;
— the annual increase in the world's food production is only half the world's population growth.

— million of tonnes of hazardous waste flood the environment every year; and,
— about a hundred species of plants and animals are predicted to disappear every day.

Based on such facts, some world leaders have claimed a plausible cause of an eventual third world war might be environmental degradation, leading to vast streams of "environmental refugees". However, there is a widespread awareness globally of the potential for such a catastrophe, and a growing but firm conviction on the part of many never to allow it to become a reality. Unfortunately, not everyone shares this conviction.

This level of environmental concern is unique in the history of planetary evolution. It may affect all of us and our possibility of surviving as a human race. Therefore, the world's population must be united in this struggle for survival. Department of environmentally-efficient technologies, participation of the individual, and a change in our thinking and behaviour should help guide us to a sustainable path to ecological stability, bearing in mind that equity and justice ought to prevail within and between human generations.

During the course of time, population-driven development has increased as well as technically-mediated interference as a part of our way of living, but human solidarity may have decreased somewhat. Our aspirations for greater material wealth have made many individuals more isolated, living wasteful lives, which has imposed a strain on the environment.

Good environmental quality has, more or less, come to be regarded as something unchangeable and obvious. The continuous degradation of environmental resources has made people more aware of the susceptibility of the biosphere to human interventions.

There is a growing awareness that environmental; problems are as much a question of life-styles as of agricultural or industrial activity. Therefore, it is necessary to change people's attitudes with regard to the environment.

A new environmental behaviour is necessary, in which quantitative demands and confrontation must be replaced by qualitative appreciation and coordination. This will hopefully lead us into a new era of human harmony where all care for the well-being of life on Earth. People can make their contributions to protecting the environment in two ways, either passively or actively.

The passive way reflects the ability of green consumers to re-orient their choice of purchases in favour of environmentally-friendly products, which will give important signals to manufacturers and retailers.

However, in their role as consumers, many people are often dependent on simple, reliable information, in the form of media coverage of environmental matters, advertising, and different forms of eco-labelling of products. Another important but passive individual contribution to a better environmental quality is separating household waste, as a means of enhancing the recycling potential.

The active way to assist in environmental protection work is through participation in environmental organizations, so-called "green parties", and in the many forms of citizen's environmental advisory boards. However, a crucial point is access to honest, understandable information on the causes and effects of environmental disturbances. Individuals can make up a valuable collective memory, essential for promoting civil authority or soliciting governmental feedback in establishing environmental protection programmes. Direct public participation in the planning and decision-making process often makes people more amendable to apparent sacrifices and associated excess costs of environmental management.

Growing environmental awareness

Mass production	→	Differentiated production
Expansion	→	Consolidation
Specialization	→	Generalization
Exploitation	→	Reuse, renew, repair
Mass communication	→	Selective information
Economy	→	Ecology
Money as values	→	Meaning as values
Lonelyness	→	Cooperation
Privacy	→	Openness

The gradual dominance of nonpoint-source pollution from a variety of human activities over the more easily-identified and controlled point sources allows no sector of society to escape its responsibility to make their individual efforts for reducing their over all environmental impact.

It calls for a new, superior form of social environmental behaviour to facilitate a process of introducing new environmentally-friendly technologies and changing life-styles-from the traditional patterns of industrial, and information and communications societies, to a new era of consciousness about safeguarding the environment in a sustainable society. Environmental issues must be fully integrated into socio-economic planning, and include a wide range of responses across a large number of public sectors-research, education, communications, industrial production, public construction, consumptions, waste management, etc. Social and ecological planning should be based on enthusiasm founded on knowledge and cooperation. All parties must feel on increased responsibility and awareness to stimulate a fruitful environmental dialogue. Every sector should be coordinated in their environmental ambitions, to create the necessary mix of responses to best suit local conditions. It is of vital importance to overcome conflicts arising from the differing view-points of the various productive economic sectors and governmental departments with regard to their budgetary means and associated possibilities for enforcing remedial actions. For example, it is not unusual that municipal environment and health departments are small units of restricted financial means. It might be necessary to establish a joint community environmental council with representatives from all sectors of society. Another alternatives may be to encourage people with broad range of environmental skills to join the Executive Boards in each of the major economic sectors.

A necessary prerequisite for carrying-out an effective environmental protection programme is gaining access to scientifically-sound and relevant information about the pollutant contributions from each sector, and how these pollutants can interact and influence the environment. A general recommendation is to make environmental impact assessments, prior to initiation of each public project. Environmental policies have to be in conformity with the overall, sectoral policies of society. However, such an ambition can easily lead to conflicts . Now might be the time to abolish the setting of a specific energy policy, and instead replace it with a sound environmental and economic policy.

Energy issues could then be handled and regulated, based on rational demands placed on environment, health and safety. Every country has a unique character. Thus, integrating environmental aspects into socio-economic planning might take time in countries that still need to satisfy basic human needs. However, the earlier environmental issues can be included into ongoing development processes, and institutional capacities strengthened to build up a new environmentally-sound infrastructure, the sooner the community will realize economic profits from well-maintained natural resources.

Burdens and advantages will be unevenly distributed between different countries until generally-agreed international environ-mental policies are established. The quest for forms and new rules governing international relationships and cooperation must be met. The developed countries should take a lead in this process, and find new forms of cooperation to mediate their relationships with the developing countries, mainly because many earlier cooperation programmes have over-exploited natural resources.

The new forms of international cooperation should include the direct involvement of developed countries to build up the scientific and technical skills of developing countries-no longer solely in the form of gifts or technical packages, but also supplemented by human relationships.

It is important to provide for full participation, integration, and empowerment of people in developing countries in order to solve their own problems. Such a process necessitates open access to all types of relevant information, as well as free expressions of views. It also necessitates exchange of technology, or at least the creation of conditions which will facilitate the exchange of technology between countries in need, in-cluding not only developing countries but also countries in transition. Poverty and water scarcity, in many cases caused by the rapid population growth, are at the root of many emerging problems in developing countries. Two of the most critical problems are degradation of natural resources and malnutrition.

Progressive environmental work in developing countries is generally restricted, because of shortages of financial, technological and human resources. Therefore, international organizations should explore all possible ways of encouraging family planning and of finding effective funding mechanisms to assure access to the same remedies used elsewhere, preferably in the form of educational and training programmers, and the transfer of the best-available technology.

At the same time, the protection of proprietary rights must be guaranteed. One should not forget that international cooperation with regard to environmental protection is often an integral part of a country's development policy. Because many developing countries exhibit a minimum of infrastructure, aid programmers often take on the form of eco-development programmes.

Reduced exploitation of natural resources should be carried out with recognition of economic, cultural and social realities. International organizations should support the efforts of developing countries as well as of countries in transition to strength their competence, and to create and develop indigenous capabilities parallel with protecting and preserving the country's cultural heritage.

Much tradition and knowledge in developing countries about ecosystems remains unrecorded and unused; this knowledge is extremely valuable for better use of our biological wealth. More attention must be paid to improving the quality of aid programmes in developing countries, to favour effective environmental work which incorporates their rich and diversified culture.

Environmental protection has a short history. In its early days, much attention was given to collecting evidence of ongoing degradation of wildlife and ecosystems. A number of environmental disasters were also linked to vast discharges of hazardous chemicals. Gradually, relationships between pollutant loads and their environmental consequences became clearer. Today, it is possible to forecast, in general terms, many environmental threats resulting from large-scale and complex human activities.

There is presently sufficient knowledge to initiate effective remedial measures for most environmental problems. As examples, many people agree that;

- there is a conceptual and technical basis for managing tropical forests, and maintaining their diversity;
- there is sufficient scientific knowledge on pollutant emission and their environmental impacts;
- there is basic experiences of environmentally-friendly industrial processes to executing a broad range of policy actions which developed these processes;
- there is the possibility of long-term energy sustainability, offered through energy technologies based on energy efficiency and conservation, as well as the introduction of renewable sources, and;
- there are many motivating methods such as public education, legal and institutional arrangements, economic market and financial incentives to achieve sustainable development.

It is generally believed that the lag in fully enforcing effective environmental protection efforts is mainly due to a communications gap between scientists and engineers, on the one hand, and between planners and decision-makers, on the other. This dilemma also leads to a lack of continuity between the theory of sustainable development, and its practical implementation in environmental protection programmes.

Much is at stake with regard to the urgent need to establish and enforce effective environmental policies. The many people involved in this process must do the proper things at the appropriate times, in a chain of events from synchronizing scientific, engineering and technical skills, to make an understandable digest of the relevant environmental information to the people in charge of enforcing legislation and remedial measures.

Authorities, in close cooperations with non-governmental organizations, are usually those in charge of devising environmental policies and environmental protection programmes. Sometimes it is suggested that environmental plans are vague and unrealistic, and developed partly in conformity to suit with public opinion.

Sometimes, environmental plans even seem to be formulated solely to satisfy the "green movement", with little real significance to accomplish relevant problem-solving activities. The new types of complex environmental problems ahead of us require new types of decisions and, hence, a new approach to the planning process.

Earlier, and in some cases still existing, environmental plans are more or less oriented to ameliorating easily-identified point sources. Due to the growing awareness that other pollutant sources have gradually become more important, the patterns of interrelated and mutual relationships between all of the different polluters over a much wider area than a single stack or pipe, is difficult to overlook.

Several new planning components must be considered, and environmental protection programmes must incorporate new time perspectives. The general characteristics of future environmental plans should be that they are;

— *relevant,* with regard to being able to fulfil established environmental objectives;

— *operational,* with regard to being cost-effective on a regional and global scale implying re-allocation of environmental investments beyond national and regional borders; and,

— *realistic,* with regard to feasibility.

These various prerequisites for future environmental planning put special demands on the people involved in the planning process. Several new environment, economy, trade, international aspects, etc,. Thus, the ideal future "planning person" would be someone with a basic scientific training and possessing experiences in environmental protection work across a number of various fields. However, such individuals are rate. Authority officials often have a lot of practical experience in environmental protection work, but might lack in-depth scientific training.

People representing environmental groups, on the other hand, often have genuine scientific training, but might lack practical experience in environmental protection work. It is important to guarantee everybody's right to be part of the planning process. Multidimensional environmental issues need additional experts representing, e.g., the business sector, labour and the public. Industrialists, technicians and economists will add necessary knowledge and valuable insights needed to produce a meaningful and workable plan. They all make up a planning team, with the goal of addressing environmental problems in their totality. An important advantage in creating such "broad" planning teams is the fact that it prohibits anyone to give priority to only their specific "territories". It also helps make possible the direct acceptance of a programme by all parties involved in the implementation plan, since all parties have been involved, up-front in its creation.

The basis of all planning activities often comprises a mix of data and information complied by various individuals representing different fields of expertise. People involved in the preparation of such background materials usually are scientists, engineers, economists and planners.

However, they sometimes have difficulties in cooperating with each other, not only in terms of combining their data sets and conclusions to produce a homogenous and understandable message for the decisions-makers, but also in terms of a lack of common "language".

The difficulties might consist both of lack of sufficient education in adjacent environmental matters, different time perspectives, different types of technical materials, and difficulties in communication.

Scientists, engineers, economists and planners often have different time perspectives. Scientists often have to review evolutionary time scales to explain the value of an endangered species, as precursor to their making any judgements on threats posed by exposure to pollutants. Engineers often judge the value and longevity of technical innovations and facilities in terms of endurance and reliability.

Economists often judge them with regard to the necessary time required for income generated to pay off investments or depreciates a capital item. Political and administrative planners, on the other hand, in some cases regard a suitable period of time coupled to mandate periods for realizing their ambitions to deal with environmental protection.

Scientists, engineers, economists and planners also work with very different aspects of the environment, often less compatible with each other. Scientists often work with probabilities, engineers with realities, and economists with financial and commercial forecasts dependent on market developments. Political and administrative planners, on the other hand, often work to realize promises and policies.

Scientists, engineers, economists and planners also seem to have difficulties communicating their information and findings to each other and to other parties. Scientists might provide an "academic" approach to a problem,. describing all of its associated uncertainties.

Many are unwilling to commit to any single, unqualified outcome. Engineers might contribute with very complicated and technical solutions and are usually very willing to propose a single, technical solution to any situation. Economists are likely to present several approaches to a problem, based on different fiscal and monetary policy assumptions, which usually boil down to cost considerations.

Political and administrative planners structural or physical planning investigations, but are often very vague about how the interdependence between social factors influenced the final outcome. While the differences described above do present (sometimes very great) problems, they are not insurmountable. More and more frequently, engineers, economists and planners are being exposed to environmental issues during their training, and some scientists are beginning to gain exposure to planning, business and engineering topics (although generally less so than their colleagues in the other disciplines).

The development of empirical models is helping scientists to provide numerical responses to planning questions (together with their calculations. The trend towards the introduction of national environmental accounting into the economic analyses is beginning to result in "true" cost accounting being applied to various technical alternatives.

As a result, better and more quantitative planning methods are beginning to better account for the human/ social factor in physical planning decisions. An effective environmental programme is dependent upon a proper basic environmental education to fully reflect and use the relevant aspects and contributions from the scientist, the engineer, the economist and the planner.

The divergent "messages" from all those parties might be due to the fact that the target audience is obscure to them or differs between disciplines. Therefore, the intended audience for the programme must be identified at an early stage. All parties involved in the planning process must pay considerable attention to expressing their views in easily understandable, and even popular, terms.

This might be a difficult task to many, who have come into their profession at a time when technocrats were rarely questioned and it usually takes additional time to develop the necessary "people skill" that is required in effective and lasting planning. Some individuals might feel that it requires them to go beyond their professional skill and honour, to popularize their input in preparation of a planning document. But it is an important to do this as it is in preparing the plan to learn all the relevant technical terms used by other parties, so that everyone's contribution can be understood. Thus, these different contributions to the planning process must be synchronized as much as possible, in terms of general points of departure, time spans, ways of formulating hypotheses, uncertainties in the prognoses, and an understandable way of presenting a digest of the most important conclusions about the plan to the public and decision-makers. This is probably something that requires several years of planning experiences to accomplish.

The only way to ensure such a synchronization between all parties involved, is the introduction of environmental matters in all forms of socio-economic planning at any early stage in the process followed by an audit procedure or post-project analysis at different levels

Many planners have the feeling there is not enough available information to permit the establishment of long-term environmental policies. However, this is not always true. First, there actually may exist too much information on environmental problems, but much less information on environmental issues which allows a quick overview and understanding of a problem. To overcome this, available information in the field of environmental protection must be digested and reduced to a suitable size and style to suit different target audiences.

The audiences, in turn, must be able to easily absorb the concepts of causal relationships between pollutant loads and ecosystem responses as well as the main conclusions on how to cope with the problem. In many cases, these individuals will forward "messages" to other groups of people, either to their operating staffs in industry for practical execution of the programmes or to the public, media and/or politicians for information and the use of decision-making.

Therefore, two main types of information can be distinguished, either being:

— detailed and practical user-oriented information; or,
— general information covering a wide field of environmental issues.

When providing detailed information, it is very important to identify the specific audience in question. Special attention should be devoted to making the message:

— short;
— relevant and factually-based;
— concentrated;
— understandable; and,
— concrete and applicable.

It is beneficial to consider no more than a couple of main messages based on a few key arguments. These should preferably be supported by some practical examples, and repeated at the end of the information package. When giving *general information*, it is important to systematically simplify the use of scientific knowledge.

With the main aim of providing relevant background material to planners, such information should be best given in a way that conforms to the questions that need to be addressed and answered in the decision-making process in other words, target the input to the question and avoid the tendency to ramble or pontificate. Such stepwise information procedure can facilitate the understanding of different types of environmental problems, and make comparisons between them easier to understand.

The increasing complexity of environmental issues will inevitably lead to greater implications to several sectors of society, irrespective of the characteristics of specific decisions and their consequences when implemented.

Many new approaches in the field of environmental protection affect a broad spectrum of the world's population, in their role of private citizens, employees, managers, authority or governmental officials, company leaders and/or politicians. Ten general examples of such future approaches are:

— widening of public participation in the planning and decision-making process;
— strengthening of market potential and role of green consumers inclusion of environmental impact assessments, that are based in general terms, on the critical load-concept;
— accelerating the need for infrastructural changes;
— adoption of the "cradle-to-grave" approach to industrial processes and products by an increasing number of industries;
— introduction of environmental or "green" taxes;
— focusing on the responsibility of polluters for giving evidence of environmentally-friendly behaviour:
— establishment of environmental quality norms, based on user aspects;
— increasing awareness and implementation of global environmental aid programmes and policies.

No one involved in establishing future environmental strategies should be able to decide and pursue a policy in one area of society, without considering its effects in another area. Responsible planners representing any sector of society would probably recognize the value of being able to forecast the social and environmental consequences of a project by considering all relevant effects on future human development. While this may be idealistic still at present, this ability is rapidly being developed in many communities. It is extremely difficult to continuously gain insights into the environmental ambitions of the various sectors of society that have general implications, when planning environmental protection programmes.

However, many of these ambitions are contained in the international literature, as well as in conference and meeting proceedings, leading to different statements and declarations. While the many difficulties in surveying all the information is recognized, the possibility of digesting and disseminating this information through electronic media may bring this goal into reality.

Community vision is articulated by scientists, legislators, people working with various environmental issues in municipalities, country administrations, government authorities, and politicians and experts taking part in international negotiations and agreements. It also includes citizen inputs and should be directed at these same citizens. In other words, it needs to be "real" and not just a philosophy. To be "real it needs to be firmly grounded on the scientific, administrative and holistic principles and considerations set out below. Scientific perspectives underlying the community vision should seek;

— to develop methodologies for identifying environments of primary concern to assist in prioritizing steps towards a practical implementation of the aim of sustainable development;
— to develop analytical procedures to further expand the priority-pollutant approach for identifying environments of primary concern to assist in prioritizing steps towards a practical implementation of the aim of sustainable development;

— to develop analytical procedures to further expand the priority-pollutant approach for identifying new hazardous to human health and the environment:

— to make further use of "biomarkers", for a more unequivocal elucidation of the "environmental fate" of pollutant emissions;

— to develop means of comparing, weighing and evaluating different types of environmental problems, with regard to their detrimental effects on biota, and their reversibility;

— to find new approaches, using "material balances", to identify the routes and magnitudes of transportation mechanisms and transformations of chemical compounds and pollutants in the technosphere and the biosphere;

— to further develop methods for the practical use of source-receptor and load response relationships for all types of environments of primary concern;

— to widen the concept of critical load analysis to cover regional aspects, and the influence of the dynamics of gradual accumulation of hazardous compounds in the environments, and the subsequent impoverishment of natural buffering systems;

— to increase the predictive power of using risk assessment and risk management, by including the concept of total human exposure; and,

— to increase efforts to find ways of using biotechnical innovations to render hazardous chemicals harmless.

Administrative perspective should address the following issues in the definition of a future community visions; administrators need;

— to introduce a new sense of social environmental education and planning, focusing on impaired closed cycles of material flows;

— to give higher priority to infrastructural changes, in order to combat deterioration of the environment following the growth in the transportation sector;

— to work out and implement "sound environmental legislation".

— to simplify the administrative procedure when reviewing, issuing and renewing pollution permits, and initiate a system of rewarding instead of charging;

— to introduce a new taxation system, putting a tax burden on resource use and pollution, instead of on labour and capital;

— to expand the right of private citizens to contest pollutant permits and other regulations resulting in harm to the environment;

— to request information, in the form of "hazardous waste declarations', from every party not capable of handling their own waste;

— to demand comprehensive investigation of minimizing environmental impacts from present waste-disposal sites, and thoroughly examining the suitability of new ones before they are taken into operation;

— to establish markets for trading hazardous wastes; and,

— to reallocate time to permit staff to enlarge the scope/breadth of their activities focusing on the establishing of guidelines and criteria, and to digest and evaluate information on monitoring programmes and long term planning to promote more informed and holistic decision-making.

International perspectives are needed to produce a holistic view of the global economy and environment.

Governments need to promote the community vision by taking action:

— to support and encourage the ongoing work by various international expert groups on environmental impact assessment and to facilitate international cooperation on environmental solutions;

— to make more efforts to provide continuous stimulus to a fruitful North-South dialogue on environmental matters, in the same manner as EAst-West environmental cooperation, which is already under way;

— to promote a world-wide exchange of environmentally favourable proposals, ideas and recommendations to facilitate the free flow of the information and technology needed to combat global environmental problems;

— to strengthen international organizations in taking active part in the transfer of modern treatment technologies to all parts of the world-internal financial mechanisms should be identified and established to help in exchanging treatment technologies as well as any environmentally-sound technology.

— to find new funding mechanisms to ease the debt service burden of developing countries;

— to consider the expansion of the "Polluter-Pays-Principle". to include every nation's responsibility for the control of transboundary pollution and the adjustment to the critical load concept;

— to support that expenditure from one country could be invested in neighbouring countries, to make more environmental use of the reducing transboundary pollution at its source, thus maximizing the benefits obtained from financial resources;

— to encourage developed countries to reconcile future development from the standpoint of ecological as well as economic objectives;

— to assure the participation of international organizations in existing database systems for environmentally-sound technologies and,

— to stimulate education to absorb new techniques and innovations for developing sustainable environmental and energy-sound future in developing countries which respects their socio-cultural and institutional characteristics.

The commercial visions might include perspectives of various experts in the fields of environment, economy and energy. it also can include those in a company working with environmental issues-from operating staff to company management. Commerce fuels the demands for exploitation of resources, but also provides the jobs, monies and impetus for future environmental actions. It is likely that the realization of this vision will drive the achievement of sustainable development and bring about the new world order needed to overcome the potentially fatal consequences of global environmental damage.

Energy perspectives consistent with sustainable development, must change their focus from a policy excess to a policy sufficiency, and must begin:

— to encourage efforts to attain energy efficiency and conservation;

— to review the pricing and tariff systems of energy utilities, introducing consideration of time-variable costs, long-term marginal costs, and the need for incentives to reduce excess consumption and promote rational energy-use;

— to revitalize and maintain efforts of energy saving;

— to assess the advantages of decentralization of

energy suppliers, aiming at energy efficiency based on end-use philosophies;

— to realize that more electric energy is inevitably needed in the short term to enhance energy efficiency, process surveillance and operate purification plants;

— to ensure that development of future energy systems proceeds in parallel to socio-economic development focusing equally on directions and objectives, rather than on pace;

— to state, as a common policy, that remedial measures in the field of energy-efficiency and conservations could be judged in parallel with remedial measures in the field of environmental protection;

— to foster sound energy systems that are properly reflected when setting state energy policies;

— to establish long-term energy action plans and formulating policies and institutional arrangements in the right direction for setting ultimate energy goals which assure adequate energy supplies, with a minimum of economic and ecological consequences; and,

— to accept that an energy future based, to a great extent, or solar energy is a challenge to the initiative and dynamics of international cooperation.

Economic perspectives need to be amended:

— to ensure that future pricing systems fully reflect social and environmental costs;

— to accept "trade-offs" between environmental and man-made capital;

— to assess whether environmental taxes replacing traditional taxation makes an economy stronger;to determine, assuming that greater confidence is placed on market mechanisms, the economic instruments that are appropriate tools for ensuring economic growth properly considers environmental concerns;

- to assure that economic instruments are used in a harmonized way, in conjunction with administrative instruments, both to allow full appreciation of environmental costs and benefits, as well as to provide continuous incentives to polluters to make further efforts to develop "cleaner" technologies;
- to motivate the advantages of using economic instruments to producers and consumers;
- to encourage efforts to increase our knowledge on how to properly value environmental assets, in order to allow their inclusion into economic budget calculations and environmental accounting analyses and statements;
- to identify new rules of environmental-resource ownership which promote sound environmental policies;
- to find effective and competitive means of borrowing, in order to reduce in-debtedness and the outflow of economic resources from Eastern European and developing countries; and,
- to establish economic policies to be tied, if not to commodity markets, to other mechanisms aimed at ensuring that environmentally-sound developments are stimulated in order to solve the international debt crisis.

Business perspective favourable to "green" companies rest upon the need:

- to supply environmentally-minded companies with the necessary budgetary resources for applying a "precautionary environmental principle" in the entire organization and in all of its activities:

- to adopt a new openness, including exchange of technology, training and information, as well as voluntary agreements with authorities on long-term strategies to further reduce environmental impacts from industrial operations;
- to adopt a proactive search for solutions based on environmental audits and to find ways of further developing best-available-techniques (BAT) for controlling and screening of new chemicals before use;
- to introduce the concept of "product stewardship" (including domestic and foreign raw material suppliers) in a system of product Quality Assurance in harmonization with international baseline standards to provide future impetus to furthering environmental production, which precedes legislation;
- to apply a business approach to selecting and determining effective targets for energy savings;
- to accept economic regulation instead of,or in parallel with, administrative regulations, if additional measures based on own initiatives are possible;
- to apply the same or better company standards in all countries in which multinational enterprises are operating;
- to remove existing barriers to trade4 and to allow market penetration of environmentally-and energy-efficient technologies to promote integration of environmentally-benign products into the market;
- to ensure full consideration of international trade in setting environmental, social and economic objectives; and.

— to strengthen the linkage between companies, branch organizations and federations of industries to agreed upon common environmental and trade policies to support bi-and multilateral agreements within the framework of tried-and-tested trade negotiating system.

Examining environmental problems in their totality in establishing and implementing stricter environmental policies is indeed a significant undertaking and largely a function of a decision-making process. All areas of concern-technology, economy, ecology, trade, legislation,etc.-ought to be included, and their interplay in the field of environmental protection must be weighed and evaluated

Although there is little experience in these matters, there is a growing awareness that important time is running out for a policy of doing nothing of this necessary integration. It is necessary to set priorities for early initiation of those activates with the largest potential for reversing environmental degradation. However, there are no shortcuts or simple guidelines on how to proceed. It is a question of comparing different remedial measures, and their anticipated environmental improvements against each other.

The potential success of introducing environmental costs in the decision-making process, and not treating environmental assets as a free commodity, has been extensively debated.

Nevertheless, even though such an approach encounters several obstacles, it seems to be the only way to develop effective, practical environmental protection measures. A process of "trial and error" in this area will probably result in the gradual refinement of suitable approaches for setting proper priorities in the field of environmental protection.

It is often convenient to deviate time into three periods; pat, present and future. From the past, our experiences are built either on wisdom, prudence, knowledge or science.

At present, our knowledge increases by following the course of events in our own surroundings, knowledge being the basis for our innovations and decisions. In the future, it should be possible to evaluate if our past prudence, knowledge and innovations were sufficiently comprehensive and enlightened to fulfill our environmental aspirations.

A retrospective look at the history of environmental protection can yield valuable insights when deciding about the time perspective that needs to be considered when establishing our safety rules for the environment in the future. To know something about the time perspectives, and the history of environmental pollution, is an important part of the knowledge needed in realizing the necessity that environmental protection must be speeded up.

From pat knowledge, it is evident that within a period of only 10 to 15 years, animal populations can become extinct, and human food can become unwholesome.

The known environmental activity of some chemical suggests that only 10 to 15 years of their use can result in a world-wide contamination, which can even have a disastrous effects on the regional and global ecosystem. From an evolutionary point of view, time perspectives illustrated by these examples, clearly show that only small safety margins exist, if any. The only way to expand these margins seems to be to increase and apply our knowledge. When undesirable ecological effects are reported, humanity normally reacts very slowly. Rules and regulations for pollution abatement are generally not established until severe ecological damage has occurred-damage which may not have affected the human population before they were willing to act. It often takes 10 to 20 years before national and international agreement regulating a specific use of chemical is amended and ratified, or for taking necessary general counter-measures of various kinds.

One important reason for this elapse of time might be lack of sufficient knowledge to develop a convincing argument in international discussions. Another reason might be that decisions on environmental aspects are often taken with greater uncertainly compared to other social areas.

It is beneficial if unanimous decisions are taken in the fight against environmental pollution. However, these decisions often demand considerable economic resources, even though they often are limited. In order to get necessary economic resources allocated, there is an urgent need to upgrade the educational level of society on environmental issues.

This will allow society to make more rapid and active political decisions, and to make appropriate economic rearrangements within society. To get the necessary environmental aspects are often taken with greater uncertainty compared to other social areas. It is beneficial if unanimous decisions are taken in the fight against environmental pollution.

However, these decisions often demand considerable economic resources, even though they often are limited. In order to get necessary economic resources allocated, there is an urgent need to upgrade the educational level of society on environmental issues. This will allow society to make more rapid and active political decisions, and to make appropriate economic rearrangements within society. To get the necessary environmental understanding not only are financial resources needed, but also time.

It takes a lot of time and effort to educate people about different social issues within the field of environmental protection. Thus, it seems that development of societal knowledge about the ecosystem and its reaction to environmental pollution is a very important measure that needs to be taken at an early stage. It does not mean avoiding concrete actions against pollution.

Rather, it means a demand for further efforts, to increase society's educational level on environmental issues, and to avoid measures that are costly and inadequate because of insufficient knowledge about properly developing and implementing control measures.

There is no single recipe for preparing people to make relevant decisions in the field of environmental protection. It requires a massive interdisciplinary training programme including people at all levels in society, e.g., school children, students, company employees, authority officials and politicians.

Scientists need basic training on practical environmental work, just as decision makers need basic ecological training. The educated efforts should be done on a world-wide basis, bearing in mind the need for different forms of teaching because of specific conditions in countries.

Proper education should to be exercized prior to implementation of remedial programmes and carried out in a step-by-step manner. Initially, basic principles of ecology should be given be given, followed by training in various practical aspects of environmental protection work. The following components have typically been included in effective programmes.

— principles of the hydrological cycle.
— mechanisms of pollutant transfer, from the microscopic to the macroscopic scale.
— self purification capacities of soils and water.
— methods of pollutant identification;
— environmental impact assessment;
— theory and prerequisites for sustainable development;
— best-available treatment technologies;
— legal perspectives on environmental protection;
— economic and administrative instruments and their validity; and,
— principles of decision-making.

2 Climatic Changes

Man is as closely related to nature as he is to himself, because he is a part of it. An outright dependence on nature has been a striking feature of man's progress through the centuries of his struggle. Since man himself is a part of a nature he is effected and in turn effects the other constituents, which are as much a part of nature as man himself.

Climate has from the very beginning regulated man in practically every aspect of life and has played a very important role in the development of civilizations all around the world. man's impact on climate began 5000 to 9000 years ago as he was able to alter the environment by of environmental resources is being eroded to meet the needs of a rapidly growing population.

A very strange and important aspect is that of the vast difference in the perspective of the developed and the developing world as far as the responsibility to own up for the dramatic increase in the pace of the Global Climate change and measures to be adopted to restrain as well as curb it. At fact that must be understood is that The Global Climate Change is global in impact, not in sources.

A problem that is absolutely impossible for either a single nation or a single area to resolve, it seems unreasonable to blame and expect only a few nations and regions to take up the responsibility for wroughting the change and then finding means of restraining and wiping it totally from the face of the Earth, as is expected by the developing world from the developed world and vice-versa. As for example, the Report of the World Resources Institute on emissions of carbon dioxide from various countries pointing to India and China as the major share holders of the responsibility came under criticism in recent times.

The questioning of the World Resources Institute figures the authenticity of research do leads one to believe that politics could be a major handle in the quest for saving the Earth from a disaster which is as imminent as tomorrow's sun if the present conditions prevail. These climate change in the wake of already depleting world resources will naturally lead to magnifying the already prevailing North-South tensions. In fact, environmental tensions are said to be both a cause and consequence of political tensions and international disputes.

Conventions

Several conferences in the recent years have taken place which have provided international policy framework to be considered when dealing with the science of the global climate change. A brief reference is being made below to the initiatives considered most relevant.

(a) Vienna Convention for the Protection of the Ozone Layer (Vienna, Austria, March 22, 1985)

This convention was signed by 20 states and the EEC at a conference convened by the UNEP. The object of the convention was the protection of human health and the environment against adverse effect resulting or likely to result from human activities which modify or are likely to modify the ozone layer.

(b) International Conference on the Assessment of the Role of CO_2 and other Green House Gases in climate variations and Associated impacts (Villach, Austria, October 9-15, 1985) and follow up workshops (Villach, Austria, September, 28, October 2, 1987 Bellagio, Italy November 9-13, 1987)

The Villach conference held with 29 countries recommended that the governments and inter governmental organisations should take into account the results of the assessment made in their environmental programmes, and should favour the increase of public information effects on the global change issues. This meet was in regard with the assessment of the presence of carbon dioxide in the atmosphere.

(c) Montreal Protocol on substances that deplete the Ozone Layer (Montreal, Canada, September 16,1987)

This protocol signed by 24 of the 46 countries attending a conference in Montreal seeks to inhibit the production, consumption and trade of ozone-depleting compounds. The compounds are divided into two 'groups: Group I (certain CFSs) and Group II (specific halons) each subject to different limitations. The protocol also distinguishes between two groups of countries, the more developed with relatively high levels of consumption of the controlled ozone depleting substances, and the developing countries with relatively low levels of consumption.

(d) Conference on the changing Atmosphere: Implications for Global Security (Toronto, Canada, June 27-30,1988)

The conference was organized at the initiative of the Canadian government and gathered together more than 300 scientists and decision makers from 45 countries and international organisations. The conference called for urgent work on an "Action Plan for the Protection of the Atmosphere." This Action Plan, complemented by national actions, had addressed the problems of climate warming, ozone layer depletion, long-range transport of toxic chemicals and acidification.

(e) Resolution of the United Nations General Assembly on Protection of Global Climate for present and future generations of mankind (Resolution 45153 of December 1988 and 441207 of December 1989)

The resolution urged governments, intergovernmental and nongovernmental organizations and scientific institutions to treat climate change as a priority issue, to undertake and promote specific, cooperative action oriented programmes and research so as to increase understanding on all sources and causes of climate change including its regional aspects and specific time frame as well as the cause and effect relationship of human activities and climate and to contribute as appropriate, with human and financial resources to protect the global climate.

(f) International Conference on Saving the Ozone Layer (London, UK, March 5-7,1989)

This conference was held at the initiative of the United Kingdom in association with UNEP, with the participation of 123 countries and the European Community. The conference was aimed at strengthening support for the Vienna convention and the Montreal Protocol and at the examination of more radical proposals than were contained in those treaties.

(g) International Conference on the Protection of the Global Atmosphere (The Hague, The Netherlands, March 11, 1989)

This conference held at the initiative of the French Prime Minister and co-sponsored by the French, Dutch, and the Norwegian governments, produced "The Hague Declaration" which called for the development within the UN framework of a new institutional authority, either by strengthening existing institutions or by creating new institutions. The declaration also called for the creation of an "Atmospheric Fund" to provide "fair and equitable assistance to compensate countries bearing and abnormal or special burden as a result of decisions taken to protect the atmosphere."

(h) First Meeting of the Parties to the Montreal Protocol on Substances that deplete the Ozone Layer (Helsinky, Finland, May 1-5,1989)

The conference issued a declaration of intent for the complete elimination of the production and use of CFCs the year 2000. The declaration also called for a transfer of technology and of financial aid to Third World countries to enable them to institute alternatives to CFCs.

(i) Summit of the Group of Seven Industrialised Countries (Paris, France, July 14-14, 1989)

By including in its final statement related to environmental issues, the Summit clearly expressed a strong interest in the environment, with particular attention being devoted to global climate change issues.

(j) International Conference on Global Warming (Noordwijk, The Netherlands, November 6- 7, 1989)

The conference issued a declaration calling for a freeze in the CO_2 emissions as soon as possible and to be set at levels to be considered by IPCC and the Second World Climate Conference.

(k) White House Conference on Science and Economics Research related to Global Change (Washington, DC, USA, April.16-18,1990)

This conference which was initiated by the US President sought to add an integrating focus for international thought on global change, by introducing the concept of "Global Stewardship". It also emphasized a new dimension of the international dialogue on global change: that economic analysis and research on broad global change policies and on the consequences of such policies must be integrated with the science of global change.

Conference on Action for a Common Future (Bergen, Norway, May 14-16 1990)

In this conference was prepared the "Bergen Ministerial Declaration of Sustainable Development in the Economic Commission for Europe Region" covering policies, objective, principles and commitments in support of sustainable development and with particular emphasis on the global aspects.

(m) London Summit (London, UK, June 1990)

This was basically an amendment of the Montreal Protocol and aimed at establishing a funding mechanism to assist the developing countries into the transition period. Developed countries agreed at London to take all practical steps to transfer technology to the developing countries.

(n) Madrid Agreement on Antarctica (Madrid, Spain, October 5,1991)

A landmark agreement on Antarctica was signed by member nations of the Antarctica Treaty which banned the exploitation for oil and other minerals on the icy continents to protect the ecosystem and also regulated the marine pollution and waste disposal. The agreement bans mining indefinitely and says this cannot be altered until at least 50 years have passed.,

(o) Earth Summit- United Nations Conference on Environment and Development (Rio de Janeiro 3-14 June, 1992.

The historic Earth Summit held from June 3-14, 1992 in Rio de Janerio was attended by over 115 heads of states or governments. The major achievement was the adoption of Agenda 21, a voluminous 800 pages document that details how countries would go about achieving sustainable development with detailed chapters on the financial principle and mechanisms involved. There are also chapter on technology transfers.

There was also a declaration known as the (Rio? Declaration) on principles on general rights and obligations on environment protection, initiated by heads of governments at the United Nations conference on Environment and Development. During the Earth Summit, success in a limited form was seen in the different treaties signed by the attending nations. Two key treaties were signed by over 150nations-the treaty on bio diversity and climate change.

An agreement signed by the countries on the statement of principles of forestry has been regarded as a major achievement and so has the setting up of the high level commission on sustainable development to monitor the implementation of Agenda 21 by the countries.

As for the funding of the cleaning up of environment, Japan during the Summit said, it will increase its development aid by 400 million a year to 1.4 billion, Germany and France promised to raise their contributions to 30.7% of their gross national products by the year 2000. Some 150 nations signed a treaty that might if carried out decreases the emissions that contribute to global warming at the 1990 level by the year 2000.

Although as Pakistan Prime Minister Nawaz Sharrif who is also the chairman of the G-77 nations, said the world will never be the same again after the conference, the conference was as giant step towards sustainable development and creation of an ecological and social environment which is safe, harmonious, and clean, it cannot be denied that some stumbling block during the conference, left the world to wonder, whether the world really will change for better after the conference.

There were serious examples of noncooperation and lack of unanimity on serious issues like Biodiversity, a treaty which was to save. Species of plants and animals from extinction which the US refused to sign, even though there was heavy pressure on it from all over the world, and also the global warming treaty to decrease the emissions that contribute to the global warming phenomenon was watering down by the U.S.

The section on atmosphere appeared no close to the solution with Saudi Arabia insisting on dropping of world non-renewable source of energy being kept out of the final text.

The Summit did not take up many relevant issues including toxic wastes, nuclear energy, over population * There was another major stumbling block at the Rio Summit, at the question of funding of Agenda 21, the ambitious checklist of environmental measures that need to be taken in the next century.

The G-77 group of developing countries has proposed 2000 A.D. as the deadline for the North to pay 0.7% of its gross domestic product as overseas development. The U.S. which now pays 0.23% of its GDP, has opposed any such deadline. There also was a lack of consensus, as to when the funding will be organized.

The G-77 wants that the funding should be done according to the wants and needs to the developing countries, as they are the ones who are suffering the percent of the global environmental crisis created by the developed nations, whereas the developed nations have asked for the IMF to regulate the funding. This was then settled over by a midway agreement of the IMF handling the funding in accordance to the committee set up for the purpose, so that the demands of both sides be met. Expectations from UNCED may not have been fully met but there is a renewed urgency and purpose and determination to work for a more equitable and environmentally sound world order. Humanity is talking about; the issues of global warming development of the world's resources, to the global information society we live in today. At the forming front of this keen interest is a large and growing group of professionals ranging from ecologists to engineers to economists to lawyers to politicians, and even royalty, all whom claim some part of the ever-growing mosaic now unfolding.

An international conference on these topics is being held somewhere in the world about every week now. Having recently attended the Gothenburg Conference "Ecology'89" which included some, of the top names in the field, one left with the feeling that nothing very much had been resolved. In fact an attempt at producing a "Gothenburg Statement", a cry of awareness to the rest of the world that might become a cornerstone in the ongoing debate, failed miserably.

In fact the only conclusion was a restatement of the original objective of the conference and a commitment to hold similar conferences in the future. The biggest problem with conference overkill is that one becomes overly familiar with the package, the headlines and the terminology surrounding the topics, without getting at the guts of the issues in questions.

While being overloaded with superficial or redundant information new original ideas are unfortunately lost. Progress fails to occur. The need for a further conference becomes indisputable. By trying to attend one conference every month one no longer is capable of finding time to think, regroup, write, exchange valuable ideas and perform analysis on the information mass that is ever expanding.

The general quality of the information decreases as this continues. Shifts in mid career often result because of boredom, lack of understanding, etc. Is this professional development? The creative component stagnates because the same people are attending the same conferences and spend most of their research time in airplanes, hotels and getting caught up on correspondence usually dealing with funding and the next conference or the like.

Unfortunately, group pressure on these globetrotters keeps them in animation the fear of becoming uniformed is too great. The fact that no progress occurred at a conference is not a signal to try to halt this syndrome. To make things worse many conferences publish proceeding which of course are inadequately edited or referred but widely indexed with lots of keywords.

This is one of the major sources of the information glut filling libraries, people's bookshelves, and employing vast numbers in the publishing and printing industries. This only leads to more conferences since no one has time to sit down and sift through what is not worth reading. Conferences are more attractive than long evenings at the library after everyone has gone home.

The enormous expenses surrounding conferences Should be a sort of natural limiting mechanism, but the number of days in a year still seems to be the controlling factor. The most important global environmental topics as chosen by a panel of about 12 world experts were as follows: human population growth, biodiversity and conservation, climate change, forest decline, hazardous wastes, land degradation, risk from biotechnology, environmentally-degrading energy production, human ignorance and fear of change, lack of adherence to facts, North-South conflicts, conflicting government policies, military destabilization and lack of democracies, perception of risk, human pathogens, urban environment, work environment and resource depletion.

No conclusions were officially made but the discussion made it obvious that 1) environmental problems arise out of both poverty AND affluence and 2) population pressure AND patterns of resource utilization combine to bring developed countries and developing countries together on a common but uncertain ground. Global awareness will keep many people running to conferences for some time in the future. Hopefully there will be some fresh minds left for the major UN Conference in Geneva in 1992, 20 years after Stockholm.

Scientific Programmes

The need for extensive research is understood by all the scientific community and to bring out solid results and methods of curbing the accelerated climate change. Many governmental and nongovernmental organizations have been set up and have made efforts in this regard. The UNEP (United Nations Environment Programme), UNDP (United Nations Development Programme) are some of the environment programmes which are making active efforts and are doing research in this field. Apart from them, Economic Commission for Europe (ECE), the European Economic Community (EEC), the European Science Foundation (ESF), The Food and Agricultural Organisation (FAO), International Social Science Council (ISSC) World Meteorological Organis-ation (WMO), The Intergovernmental Oceano-graphic Commission (IOC) this a part of UNESCO, the Scientific Committee on Ocean Research (SCOR) and many other such organisations. Apart from them there are other scientific activities underway which are funded by different organizations such as the World Weather Watch (WWW) World Climate Research Programme (WCRP), World Climate Programme (WCP), World Climate Impact Studies Programme (WCIP), Past Global Changes (PAGES), Integrated Global Ocean,Station System (IGOSS), Human Dimension of Global Change (HDGG), Global Environment Monitoring System (GEMS), Global

Change and Terrestrial Eco-System (GCTE), European Programme on Climatology and Natural Hazards (EPOCH) and many such activities which have really helped in our understanding the intensity of the accelerated Climate Change and ways and methods to work in our fight against the imminent disaster facing our planet. Humanity is conducting an unintended, uncontrolled, globally pervasive experiment whose ultimate consequences could be second only to nuclear war.

The Earth's atmosphere is being changed at an unprecedented rate by pollutants resulting from human activities, inefficient and wasteful fossil fuel use and the effects of rapid population growth in many regions. These changes are already having harmful consequences over many parts of the globe.

This analogy between the consequences of a nuclear war was not made idealistic, scientifically innocent environmentalists, but by the more than 300 policy makers and scientists from 46 countries, United Nations Organisations, other International bodies, and nongovernmental organisations who attended a major international conference sponsored by the government of Canada, which came out with a statement to the effect that it is clearly without our power not only to alter the planet beyond comprehension within a few hours by using nuclear weapons, but also within a few decades by destroying the Earth's life support systems and radically changing the climate by contamination of the air and water with the residuals of production. Global climate is changing because of the build up in the atmosphere of Carbon dioxide, methane, nitrous oxide, the CFCs (powerful greenhouse gases as well as destroyers of stratospheric ozone), and other greenhouse gases produced by fossil fuel burning, by deforestation and by producing food for the rapidly increasing population at a global level. Some of the major issues involved in the study of the Global Climate Change are the depletion of the ozone layer, the greenhouse effect in total, acid rain, deforestation i especially in the tropical regions, biodiversity, floods and many of these kinds of related issues which are directly or indirectly involved in bringing out a climate change of such a high magnitude.The composition of the Earth's atmosphere is changing.

Detailed background atmospheric concentration measurement combined with analyses of ancient air trapped in Antarctica and Greenland ice now give a compelling picture, and also of major changes that have occurred in preindustrial times. Mounting evidence that the atmosphere is changing has increased the urgency of understanding the processes that control atmospheric composition and the significance of the changes, that are taking place. The first analysis of the effect of increasing carbondioxide concentration on global warming was conducted by the Swedish Chemist, Svante Arrhennius. Arrhennius, concerned about the rapidly increasing rate of fossil-fuel use in Europe recognised that the resulting increase in the atmospheric concentration of carbondioxide would alter the balance of the atmosphere.

Using a one dimensional model, Arrhennius estimated that if the atmospheric concentration of CO_2 doubled, the surface of.the planet would warm by about 50°C. After Arrhennious many studies were done in order to understand the reasons for the changing global climate and its future repercussions.

In 1979, a study by the US National Academy of Sciences (NAS) evaluated the impact on global climate of doubling the concentration of CO_2 relative to the preindustrial atmosphere. Diskinson also in his work in

1986 suggests that the effects of a greenhouse gas buildup radiatively equivalent to doubling the preindustrial concentration of CO_2 might warm the planet to a great extent than had previously been expected.

Apart from the studies of the Climate effects of greenhouse gas buildup there have been various studies to the effect of the future Carbondioxide emissions. By the mid-seventies, measurements of atmospheric CO_2 concentrations at Manua Loa begun by Keeling during the International Geophysical Year (1967-1958) provided indisputable evidence of a long-term increasing trend while the oil embargo of 1973 and the nuclear power debate focused attention of future energy supplies.

Thus from them on, many studies were done to the understanding of the subject, like the one done on future energy use conducted by the International Institute for Applied Systems Analysis. In 1983 two economists Edmonds and Reilley developed a detailed partial equilibrium model to investigate the effects of alternative energy policies and their implications for future C02 emissions.

In 1983 the National Academy of Sciences completed a Congressionally mandated study to evaluate, among other things the effects of fossil-fuel development activities authorised by the Energy Security Act of 1980. In the few years a number of analysis have investigated the combined effects on global surface temperature of a buildup of CO_2 and other trace gases.

The studies related to this effect have been performed by Lacis, Seidel and Keyes, Ramnathan. Dickinson and Cicerone, Mintzer. More recently, Rotman have performed their study in 1988 in the understanding of the greenhouse warming. All these studies have tried to put before humanity the actual extent and the magnitude of the global climate change which is occurring at such a rapid race, and which is having so many extra side effects, on the human life.

From various studies performed, with various methods as a medium, the magnitude of the climate change has been assessed. Not only will there be a direct change of the climate, but also the magnitude of he effects of that climate change.

The consequence of the global heating which would result if the present release rate of the major greenhouse gases are maintained for only a few decades more, would be catastrophic if our present scientific understanding is even approximately correct, and the resulting climatic change would be irreversible. It is now known that a warming of several degrees, greater than previously experienced in human history, could now occur in a few decades a time which is short compared with the life-span of a tree or a man.

Major changes will result in ecological, economic and social systems, as all are in delicate balance with the environments which in turn are dependent on the climate. There is scientific consensus that the increase in greenhouse gas emissions will result in climate change.

The Council on Environmental Quality, concluded in 1981, that the potential long-term risks of social disruption caused by increased atmospheric concentrations of CO_2 (aside from the other greenhouse gases) are significant. However, considerable uncertainty exists with regard to the ultimate magnitude of the warming, its timing, and the regional patterns of change. In addition, there is great uncertainty about changes in climate variability and regional impacts, No one can describe the precise nature of these changes in part, because of the technical demands of climate modelling and in part because of the impossibility of predicting the choices which will be made within the next fifty or so years, but change there will be. Although the crystal ball is too cloudy to reveal the details, the general course of climate change is well understood. If we allow emissions of CO_2 and the other greenhouse gases to continue unabated, the earth will warm five to ten times faster than it did during the retreat of the last ice age. Many trees cannot migrate much faster than they did as the last ice age. By the time a tree matures enough to produce seeds, the climate will be unfavourable for those seeds to take hold and produce the next generation.

The greenhouse effect, is a natural phenomenon that plays a central role, in determining the Earth's climate. The hot surface of the sun radiates as a body having the equivalent temperature of 5800 K. The bulk of the radiation is in the visible wavelength region, 0.4-1.0 m, where the earth's atmospheric gases absorb only weakly. In contrast, the low temperature earth emits radiation to infrared wavelengths for the atmosphere is highly absorbing. In an oversimplified description, the atmosphere lets shorter-wavelength radiation in, but does not let the larger-wave length radiation out.

By supposed analogy with the behaviour of panes of glass, the effect is called the green house effect'. In reality greenhouses do not depend on trapping radiation for their warmth, but rather inihibit cooling of the plants due to moving air, thus, thin plastic works about just as well as glass. Although technological advancement in industry and agriculture, have provided extraordinary wealth to a portion of the global population of over 5 billion people, these technologies have the potential to dramatically alter the Earth's climate by causing changes in the composition of the atmosphere. Myriad human activities and continued population and economic growth raises the prospect of accelerated greenhouse gas building in the future. Global increases in the atmosphere concentrations of Carbon diodixe, nitrous oxide, methane and chloroflouro carbons are now well documented, perhaps already committing the Earth to significant climate change. In addition to all these changes, tropospheric and statospheric chemistry are being modified due to the addition of these gases as well as emission of carbon monoxide, nitrogen oxides, and other compound. The increase of the different greenhouse gases have been documented by the United State Environmental Protection Agency, Office of Policy, Planning and Evaluation in 1989. The concentration of carbon dioxide into the atmosphere has increased by 25% since the industrial revolution.

Detailed measurements since 1958 show an increase of about 35 parts per million by volume. Both land clearing and fossil fuel combustion have contributed to this rise, but the fossil fuel has dominated in recent years. Carbondioxide is increasing at a rate of about 0.4% per year and is responsible for about half of the current increases in the greenhouse effect. The concentration of methane has more than doubled during the last three centuries.

There is considerable uncertainty about the total emissions from specific sources of methane, but the observed increase is probably due to increases in a number of sources, as well as changes in tropospheric chemistry. Agricultural sources, particularly rice cultivation and animal husbandry, have probably been the most significant contributors to historical increase in concentrations.

But there is the potential for rapid growth in emissions from landfills, coal seams permafrost, natural gas exploration and pipeline leakage, and biomass burning associated with forest clearings in the future. Methane is increasing at a rate of 1% per year and is responsible for about 20% of the current increases in the greenhouse effect. The concentration of nitrous oxide has increased by 510% since preindustrial times.

The cause of this increase is highly uncertain, but it has been understood that the use of nitrogenous fertilizer, land clearing, biomass burning, and fossil fuel combustion have all contributed. Nitrous oxide is over 200 times more powerful, on a per molecule basis, than carbondioxide as a greenhouse gas, and can also contribute to stratospheric ozone depletion.

Nitrous oxide is currently increasing at a rate of about 0.25% per year, which represents and imbalance between sources and sinks of about 30% Nitrous oxide is responsible of about 6% of the current increases in the greenhouse effect. CFCs were introduced into the atmosphere for the first time during the century, the most common species are CFC-12 and CFC-11. which had atmospheric concentrations in 1985 of 380 and 220 parts per trillion by volume respectively.

While these concentrations, are tiny compare with that of carbondioxide, these compounds are about 30,000 times more powerful, on a per molecule basis, than carbondioxide as a greenhouse gas and are increasing very rapidly 5% per year from 1978 to 1983. Of major concern because of their potential to deplete stratospheric ozone, the CFCs also represent about 15% of the current increases in the greenhouse effect.

The chemistry of the atmosphere is changing due to emission of carbon monoxide, nitrogen oxides, and volatile organic compounds, among other species, in addition to the changes in the greenhouse gases already described. This alters the amount and distribution of ozone and the oxidizing power of the atmosphere, which changes the lifetimes of methane and other greenhouse gases.

Changes in global ozone are quite uncertain, and may have contributed to an increase or decrease in the warming commitment during the last decade. An important relationship exists between the causes of the global greenhouse effect and the depletion of stratospheric ozone: the key greenhouse gases fluorocarbons- 12 and- 11, nitrous oxide, and methane also affect Strodospheric Ozone strongly.

These gases have atmospheric survival times, long enough to permit them to make the upward journey into the stratospheric ozone layer. In the harsh ultraviolet light there, the fluorocarbons are decomposed, yielding chlorine atoms that destroy ozone. Nitrous oxide decomposition yields nitric oxide that act similarly. Methane acts to slow the attack on ozone, but it also decomposes to yield molecular fragments that are involved in ozone destruction.

Interestingly, present models say that chlorine attack on ozone is slowed, somewhat if in the future nitrous oxide and methane continue to increase along with the fluorocarbons. The slower decrease of total ozone will result in a redistribution of ozone towards lower altitudes. Carbondioxide is not involved chemically but once in the upper atmosphere it acts to cool the air by radiating energy to space. This cooling should also act to slow the chlorine attack on ozone little. Another connection arises because stratospheric ozone itself can affect temperatures at the surface of the Earth. If, for example, the fluorocarbons, nitrous oxide, and methane all increase in the future there should be more ozone present in the lower stratosphere and a warming should occur.

More complicated phenomenon may be occurring over the Antarctica, where springtime ozone concentrations have decreased greatly in the past fifteen years or so. The same ozone destroying chemicals just mentioned are probably at play but through a more complex pathways. Of all the gases relevant today, the fluorocarbons stand out in several ways. Their concentrations in the atmosphere are the smallest, the annual percentage rates of increases are the largest, and their sources are the simplest to understand; the manmade.

What is so terrible about the depletion of ozone layer is the fact that the increased levels of ultraviolet radiation, while the stratospheric ozone shield thins, will cause a significant rise in the occurrence of skin cancer and eye damage and will be harmful to many biological species. Each 1% decline in ozone so expected to cause a 4 to 6% increase in certain kinds of skin cancer. A particular concern is the possible combined effects on unmanaged ecosystems of both increased ultraviolet radiation and climate changes.

Over the last decade, a decline of 3% in the ozone layer has occurred at mid latitudes in the southern Hemisphere, possibly accompanying the appearance of the Antarctic ozone hole; although there is more meteorological variability, there are indications that a smaller decline has occurred in the Northern Menisphere. Changes of the ozone layer will also change the climate and circulation of the atmosphere. Although the phenomenon of "acid rain" (more acid deposition) was identified in Manchester, England, as long ago as 1852, and described more thoroughly in 1872, modern scientific research has been going on only since the mid-1950s. Public concern about the problem began in the late 1960s, and arose in North America 10 years ago. Acid precipitation is a mixture of strong mineral acids sulphuric, nitric and in some locations, hydrochloric in rain and snow, and is usually defined as having a pH of less than 5.6, the value of distilled water in equilibrium with atmospheric carbondioxide.

The primary reason for concern is that acid deposition acidifies streams, and taken on coarse, sandy soils low in lime. The effect is seen particularly in heavywater areas and in wet montane environments, wherever sulphate loading from anthropogenic sources is strong.

In Norther England, several lakes on soils poor in lime were very acid by the 1950s. In southern Scandanvaia thousands of lakes and streams in the Adirondacks, Maine, Ontario and Nova Scotia have likewise shown acid stress in recent years and thousands more are threatened, particularly in Canada.

The chemical and physical consequences of lake acidification include, increased leaching of calcium from terrestrial soils, mobalisation of heavy metals such as aluminium, zinc, and manganese and an increase in the transparency of lakewaters. The biological consequences include market changes in communities of aquatic plants and animals, with a progressive lessening of their diversity.

In a few extreme cases a lessening of primary productivity is observed, and there is a progressive elimination of sensitive species of plants and animals, including fish which are completely exterminated under severe acidification. Many ecologists, believe that over the long term, several decades to few centuries—acid deposition may further impoverish forests soils, developed on sandy substrata poor in lime. As a consequence of accelerated leaching of nutrients, such as phosphorous, potassium, magnesium and calcium from these soils, forest productivity would eventually be reduced.

Acid precipitation can also mobilize heavy metals from pipes into supplies of potable water. Some ground waters have known to become acidified in Sweden. Moreover, the acid sulphate particles that contribute to acid precipitation are in the size range that penetrates deep into the lung, and they may well exacerbate lung diseases and increase mortality rates. Forests, which store 20-100 times more carbon per unit area that croplands, play a critical role in the terrestrial carbon cycle. Each year at least 11.3 million (and perhaps as high as 15 million) hectares of forest are cleared in the tropics.

The rate of deforestations combined with the exalting growth in demand for forest products is such that while 33 tropical countries are currently net exporters of wood products, this may decline to fewer than 10 by the end of the century. If this trend could be halted and reversed, tropical forests could serve as a vast carbon sink, reducing CO_2 levels.

The causes of deforestation are well known. They include population pressure for agricultural land, the demand for industrial timber production and export, and inappropriate government policies regarding land tenure, economic incentives, forest settlement, and other population issues. In tropical Africa and in South and Southeast Asia, rapid population growth appears to be the critical factor affecting deforestation. The majority of the population practices agriculture, and most of the increases in agricultural production necessary to sustain high birth ratio have come increases in the area under cultivation through deforestation. The Amazon region in Brazil is experiencing one of the highest rates of tropical deforestation in the world.

Agriculture

Agriculture contributes to the emission of greenhouses gases through three primary means rice cultivation, nitrogenous fertilizer use, and enteric fermentation in domestic animals.

Estimates place the annual contribution of the cultivation and domestic animals at approximately 20 and 15%, respectively of global methane production. Both the magnitude of agricultural source emissions themselves and the potential effectiveness and costs of possible reduction measures are very uncertain.

While considerable research has been done on the agricultural activities of interest, relatively little attention has been focused on agricultural related emissions of greenhouse gases and how various changes in agricultural practices affect these emissions. It has been found out the emission from all three categories of agricultural practices are expected to increase over 1985 levels: Global CH_4 emissions from rice and enteric fermentation increase about 35% and 65% respectively, by 20-25; and N2O emission from fertilizer use are projected to increase by 133% by 2025. By 2100, emissions from rice, enteric fermentation, and nitrogenous fertilizer increase by approximately 40%, 125% and 175%, respectively. Previous natural climate changes have cause large-scale geographical shifts, changes in species composition and extinction among biological communities. If he widely predicted greenhouse effect occurs, communities would respond in similar ways.

Global warming

Population reduction and habitat destruction due to other human activities would make it difficult for species to shift ranges in response to changing climate conditions. Human encroachment and climate change could jointly threaten many more species than either alone. It can be inferred, how the biota might respond by observing the world as it is to day. By observing present distributions of plants and animals, which are determined in large part by temperature and moisture patterns, it is possible to hypothesize what would happen if these underlying temperature and moisture changes.

For example, if it is known that one race of the dwarf birch, 'Betiulla nana' can only grow where the temperature never exceeds 22°C, then a hypothesis could be made, that it would disappear from those areas where global warning causes temperatures to exceed 22°C. Observations have led to believe, that, plants and animals are very sensitive to climate. Their range move when the climate patterns change-species die out in areas where they were once found and colonize new areas where the climate becomes newly suitable. It is also known from the fossil records that some species have become\ completely extinct because they were unable to find suitable habitat when climate change made their old homes unlivable. Geological stress would not be caused by temperature rise alone.

Changes in global temperature patterns would triggers widespread alteration in rainfall patterns, and for many species, precipitation is a more important determinant of survival than temperature. Some regions would see dramatic increases in rainfall, and others would lose their present vegetation because of drought. Other environmental facts would change because of global warming soil chemistry would change.

Increased carbondioxide concentration may accelerate the growth of some plants at the expense of others, possibly destablizing natural ecosystems. And rises in sea level may inundate coastal biological communities'. One important pattern of global warming, generally concluded by a variety of computer projections, is that warming will be relatively greater at higher latitudes. This suggests that although tropical systems may be more diverse and are currently under greater threat because of habitat destruction, temperature zone and arctic species may ultimately be in greater jeopardy from climate change.

Also a recent attempt to map, climate-induced changes, in world biotic communities projects, that high-altitude communicates communities could he particularly stressed. Boreal forest for example was projected to decrease by 37% in response to warming of 3°C. An issue that was heavily debated in the recently held UNCED in Rio de Janerio, Biodiversity is one of the major problems that the world faces today, that of either a heavy reduction in some of the plant and animal species, or a total extinction of some other plant and animal species, which cannot adapt to this rapid change.

All the issues discussed and many others like them will bring about many drastic changes on the face of the earth. They would impril human health and well being, diminish global food security, through increases in soil erosion and greater shifts and uncertainties in agricultural production, particularly for many vulnerable regions.They would also change the distribution and seasonal availability of fresh water resources, increase political instability and the potential for international conflict, jeopardize prospects for sustainable development and the reduction of poverty, accelerate the extinction of animal and plant species upon which human survival depends, alter yield productivity, and biological diversity of natural and managed ecosystems, particularly forests.

If rapid action is not taken now by the countries of the world, these problems will become progressively more serious, more difficult to reverse and more costly to address. The potential in developing countries for large percentage increases in emissions of greenhouse gases (GHG) over the next decades has resulted in suggestions that control over such emissions should be considered as part of near term development planning.

Sometimes the hope is expressed that such restrictions can be part of a win-win strategy, i.e. that concerns over global warming only lead us to take on programs such as reforestation, energy efficiency, and fuel switching, for which there are many other good reasons to undertake anyway. While undeniably true in many circumstances, this begs a crucial questions: does the threat of climate change through GHG emissions justify efforts to alter the development pathway now being followed in poor countries'?

Many observers apparently believe that the implications of global warming are so grave that, in some fashion, restrictions on GHG emissions shouldbe considered along with the other numerous and already burdensome restrictions on development, e.g. lack of capital, lack of skilled personnel, and lack of infrastructure. From the standpoint of the poorest two-thirds of the population in developing countries, concerns about climate change probably justify a policy response in which not only do GHG restrictions have no place, but GHG emissions rates may actually be stimulated locally. The best way to protect the poorest groups is to bring them as soon as possible to a minimum level of development at which they can be expected to adequately handle environmental stress.

3 Marine Resources

India has world's one fifth marine area with rich fauna. However till now the major fishing area is confined to an offshore belt of 1 sq. km. only, and the rest of the Exclusive Economic Zonel (EEZ) is untouched, which abounds in fishes of lage variety. Also, the 1981 estimate of India's offshore oil reserves were around 1100 million tonnes and the gas reserves being 12,000 ´ 10^9 cubic feet. But even then she is notable to meet her 50% of domestic requirement.

Lack of coherent maritime policy, unplanned ocean management system, inherent technological backwardness, in addition to her comparatively new awakening towards ocean resources and maritime trade, are perhaps the main causes for her being underdeveloped in this field.

A General Account Growing attention towards exploration of oil and gas, particularly after 1973 world oil crisis, mostly in the continental shelf region, and a distinct shift towards mineral exploration and exploitation in whole of the EEZ after the final declaration of UNCLOS III in 1982, other than regular search for non-conventional sources of energy within the said region represent the overall marine activities taking place in the fields of scientific and economic development taking place in respective EEZ. The main reason behind the enhancing research programs by both developing and developed nation is the increasing demand of mineral requirements by modern societies, in an almost undefined and unpredictable volume, supplemented with the fact that land resources are either on the verge of depletion level or otherwise their vital importance has been realized as a reserve in near future. Found mostly beyond EEZ, manganese nodules and polymetallic sulphides are of special value, as they consist of strategic metals, viz., cobalt, platinum, chromium etc., in a concentrated form. Classification of nonliving resources have been made on various basis. In the continental margin area unconsolidated sedimentary deposits include sand and gravel, calcium shells, placers, beaded phosphorites, metalliferous or brine rich sediments, while barities and phosphorites form the major modular and crust deposits. Selective sorting of mineral grains occurs by mechanical winnowing of wind, waves and currents, and by gravity setting and concentration. Among the selected production figures for sand and gravel from marine environment, UK and Japan top the list by producing more then 165,000,000 tons, having value more than $330,000,000 (USD) and 84,700,000 tons having value $170,000,000 (USD) respectively, mostly from a depth ranging between 35m-50m.

Calcareous sands, shells and coral sands are of course renewable biological resources, but they need special care while exploited at a commercial scale, because of their direct effects on the coastal marine environment. Even offshore bank exploitation may have an impact on the beach amenities. Found in nearshore zones, placer deposits are mechanically concentrated heavy detrital materials which originate from eroded offshore rocks, often discharged by river work in considerable amount, in the form of black or dark stringers and lenses on modern beaches.

Past sea level changes, recurrent wave actions and dominant tidal currents have however produced major locational changes, including their nature and depth of occurrence. While buried placers are little known, cassiterite exploration in Southeast Asian tin belts undoubtedly will reveal several examples, although this still remains a fact that general placers are richer on modern beaches than on submerged ones. Further wider shelves have greater number of placer deposits. A comprehensive account of the United States' EEZ placer deposits was recently given by Beauchamp and Cruickshank including its comprehensive economic futurology.

In India, National Institute of Oceanography has taken the sole responsibility of mapping, exploring and exploiting these huge deposits, mostly available on the western coast of India, in the form of tin placers, complex heavy minerals, precious metal placers and rare diamond accumulations. Besides biotite, biotite-muscovite, and cassiterite, the tin placer deposits contain pyrite, ilmenite, zircon, tourmaline and monazite, commonly found in India, Burma, Thailand, Malaysia, Indonesia, Cornwall (UK) etc.Among the complex heavy mineral sands containing principally ilmenite, rutile, zircon, monazite, garnet, and magnetite, the iron-titaniferous sands have been identified at commercially exploitable level at Southern Japan, East and West Australia, California, Oregon, Alaska, New Zealand, India, Bangladesh and Sri Lanka mainly, besides USSR, Poland, Mozambique, Italy, Philippines, Argentina, Mexico and Brazil.

As regards monazite, mineral placers were identified especially offshore India and Sri Lanka. Preçious metal placers containing gold, platinum and related metals have been mainly found concentrated in Alaska, Oregon, California, Nova Scotia, Lapter Sea (USSR), Southeast

Australia, Bering Sea etc. Diamond accumulations containing mainly kimberlite, epidote, ilmenite, magnetite, and pyrite has its only known commercially exploitable deposit at Hottentot Bay near Orange River in Namibia. Beaded phosphorite minerals mainly in the form of carbonate fluorapatite are found in the upwelling areas like Peru-Chile slope, North and Southwest Africa, Western Australia, and in the continental shelf areas of Congo, Baja California.

Metalliferous and brine-rich sediments, known in sufficient concentration in Red Sea Spreading Center, contains mineral assemblages like clays, hydrous oxides, carbonate and sulphides. However the most important brine deposits are found in the Atlantis II Deep, where economic estimates suggest about 3 ´ 10^6 tons of metals, excluding iron which forms almost 29% of the deposits. Nodule and crust deposits have been reported present, although in less exploitable quantity off southern India and in the East Indies region.

Phosphorite nodules and crusts, generally present on submerged marine plateaus and continental rises viz. Blake Plateau, the California Banks, Agulhas Bank, the Chatham Rise, and in some scattered seamounts. Known manganese nodules deposits of both mononucleate and polynucleate types are most abundant in deep seas of Pacific (in Equatorial North Pacific nodule belt, Southwestern Pacific Basin, Peru Basin, Chile Basin, South Tasman Basin and Northern Tahiti Basin), Indian Ocean (South Equatorial basin, Crozet Basin, Central Indian Basin and Western Australian Basin) and in Atlantic Ocean (in Western Atlantic Basin, Drake Passage, Western South Atlantic and Blake Plateau). Much interest is presently focussed on occurrences of cobalt rich ferromanganese crusts on seamounts and oceanic crusts, some of which have already been discovered within the EEZ of USA. Massive deposits of polymetallic sulphides have attracted recent attention. They are generally found at mid-oceanic rideges viz. stratiform sulphide bodies at Atlantis II Deep, Red Sea, sulphide mounds surmounted by chimneys at Eastern Pacific deposits and coalesced mounds surmounted by chimneys at Galapagos Spreading Center with much concentrated quantities of zinc (Zn.), copper (Cu), and gold (Au) besides some traces of arsenic and cadmium. Although exploration in Pacific have yielded much regarding their location and quantity, the possible existence in Indian Ocean, that too in considerably exploitable amount, is yet to be tested and quantified in economic terms.

In India, much emphasis has been there during the Sixth and the Seventh Five Year Plan regarding the surveys to locate and map the availability of living and nonliving resources of the Exclusive Economic Zone and that of the continental shelf. Different governmental agencies viz. Department of Ocean Development (DOD), National Institute of Oceanography (NIO), Oil and Natural Gas Commission (ONGC), Oil India Limited (OIL) etc., with their maximum emphasis on fisheries and hydrocarbon resources, are burdened with solving the problem of energy requirement of this world's second most populated country. Further, the northern Indian Ocean is one of the few oceanic regions of the world which comes under the influence of the monsoon. The uniqueness of this region, therefore, lies in the seasonal and interannual variation that it exhibits. Little is known about the nature of this variability and lesser still, about the causes behind it. Rigorous efforts are being made by sophisticated research vessels like ORV Sagar Kanya and FORV Sagar Sampada. Their utilization for some years has been going up to 82.6%. Besides, other individual and collective efforts by scientists all over the country is directed in a multidisciplinary format. The survey of the entire EEZ, according to DOD, is expected to continue for a decade or so. Intensive surveys are being undertaken to map out the nonliving resources especially minerals and hydrocarbons. Estimation of potential fisheries resources and its maximum sustainable yield is being made. The important considerations which need immediate attention from different governmental and nongovernmental agencies regarding India's EEZ are:

1. Clarification of the level up to which present EEZ utilization is going on;
2. Demarcation of optimum limit of EEZ utilization both in terms of exploration and exploitation;
3. Projected need of marine substances both living and nonliving from the EEZ by 2000 A.D., by a realistic evaluation of both protein requirements and energy substitution in different sectors;
4. Technological requirement for the fullest exploitation of EEZ resources and extent of foreign collaboration or technology transfer to be required by 2000 A.d. for its substainable exploitation;
5. Fixation of priority level, to be given to the frontier areas of ocean development in India and share of EEZ management in that, particularly in the fields of food (protein), minerals. transportation, recreation and energy, all aiming at improving India's socio-economic conditions;
6. Planning of offshore construction and required operational structure in total and the investment to be incurred.

Nonliving Resources of India's EEZ

(A) *Freshwater from sea* : The desalination technologies under consideration by the concerned governmental agencies are (1) by solar stills, (2) by flash distillation, (3) electrodialysis, (4) rever se osmosis, the last being most widely used, generating capacity of plant generally varying anything between 10,000 to 15,000 litres per day, most useful for the dry coastal areas of Indian Peninsula, presently drinking brackish water with salinity ranging from 5 to 20% in most of its rural areas. Central Salt and Marine Chemicals Research Institute, Bhavnagar has already done necessary research and development work and has made available the technology of "reverse osmosis" to BHEL, for setting up desalinization plants. One such pilot plant has been recently installed at Puthagram, near Madras.

Work on the second plant with sáme capacity of 50,000 liters potable water per day is continuing and is expected to be installed in the East Godavari district. In addition of salt, the system also remove harmful matters, like bacteria from brackish water. Besides it saves much energy as compared to the conventional thermal desalination system, thus being economically viable alternative. Although a serious threat to future large scale desalination program at global level, particularly for the vast desert areas along the oceans, is from the comparatively cheaper huge icebergs to be fetched from southern oceans, although ita waits further technological advancement and a major breakthrough in the understanding of the mysterious characteristics of icebergs, particularly the dangers to be encountered by its cleavage in any phase of its displacements. However, for localized use in coastal areas, desalination process is still economical and rather indispensable.

(B) *Drugs from the sea:* Although this utilization of marine resource is of much recent origin, by now at least 100 plants and small marine organisms have been recognized for the manufacture of important lifesaving drugs in India. More than forty are of special value, some of them containing remarkable antifertility properties, viz. Glacilaria species, with sufficient prostaglandins. At present a great need is there in India to set separate shell for studying and culturing these marine microorganisms, both of floral and faunal nature, and to help successfully in deriving and synthesizing the bioactive substances which are useful in curing many incurable diseases. Studies in India clearly suggest that all the species of marine algae, those exhibit antifertility properties may also contain prostaglandins. Their multiple use and suitability to human health is being further researched, but the most important aspect still lies with the fact that their potential presence either in the form of deposition or dissolution is still not economically beneficial for large scale economic exploitation with the existing state of technology, understanding of distribution pattern and density.

(C) *Marine chemicals:* Lesser quantitative concentration and difficult extraction methods are the major hurdles in deriving most of the chemicals economically, about 60, from the seawater. Commercial exploitation as yet has been successfully managed only for six chemicals viz. sodium, chlorine, magnesium, bromine, calcium and sulphur. Greater possibility is also there in future for deriving potassium, iodine, uranium and gold from seawater. However, optimistic proposals supported by major research findings in this field leaves us with the only hope that by 2000 A.D. India will be able to make use of its EEZ waters, mainly coastal, in deriving at least the rare chemicals which are dissolved in seawater and the same time are not easily available on land, provided concerted effort is being made to develop sufficient technological capability to meet the requirements of the complex stages involved in this process. Besides some marine organisms concentrate trace elements in their bodies by a factor of 10,000 to 1,000,000.

Cultivation and harvest of such organisms could perhaps provide a way of concentrating trace metals dissolved in seawater so that they can be commercially recoverable by biochemical means. Seventy-one per cent of the total solar energy i.e. about 106 watts reaching the earth is absorbed by oceans, thus blessing the later with the huge/energy potential.

Apart from being renewable and continues the added attraction of ocean energy forms lies in their being pollution free, resulting in remarkable minimization of health hazards, and the great competence for fulfilling the local energy requirements particularly offshore. The leading countries in this field of research are USA, Japan, France, UK and the Netherlands, implementing mostly the total systems approach, successfully overcoming the attached disadvantages. Costal inlets are best for tidal energy resources.

Technology for the utilization of tidal powers in all possible physical situations is already being developed. La Rance, since 1967, has been successfully supplying 240 MW of power to the French electric grid. In the same way, nearshore tropical and subtropical oceans are best for vertical temperature gradient, thus for Ocean Thermal Energy Conversion (OTEC). Areas of strong persistent currents, such as most of the western coastal boundaries of the world oceans are almost suitable for extraction of energy from ocean currents. River outlets are most suitable for extraction of the salinity gradient energy. Similarly coastal areas with large persistent waves can contribute maximum wave energy while wind turbines located in artificial or natural islands, near the shore, can generate maximum wind energy. In India, research and development activities are going on a large scale regarding the derivation of above mentioned forms of ocean energy.

However the forms of energy which has already been derived or could , be derived in the very near future with sufficient technological feasibility are OTEC, wave energy, and tidal energy only.

(D) *OTEC:* OTEC utilizes the temperature difference which exists between the surface seawater (28-30°C) and the water occurring at a depth of 800- 1000 metres (5-7°C) : to generate power through the medium of a suitable fluid which vaporizes at high temperature and condenses at low temperature and in this process turns a special type of turbine. According to feasibility studies OTEC power is economically competitive with fossil and nuclear powers. Directives establish a US national goal of 10,000 MW on line by 1999, from OTEC facilities only. Some of the best potential sites are located near the islands of Lakshwadweep and Andaman and Nicobar Islands groups. Only two islands, namely Kavaratti and Minicoy were found to offer possibilities for installing on OTEC plant, preferably each of I MW potential. From the available bathymetry and temperature profiles around the islands, it has been found that a depth of 1000 metres will be available within the distance of about 3 kin from the islands and a temperature difference of 20°C between the surface and 1000 in depth throughout the year.

Therefore, it has been considered that a shore based plant is ideally suited for these island sites, which also have big lagoons on western side. These islands are of a coral origin and detailed studies for placing the cold water pipeline on the sea-shope are necessary. Kavaratti was found relatively more uniform. Besides, in the technological field, procurement of metal components viz. plate heat exchangers, cold water pipes, turbogenerators, pumps, etc. are most vital for successful exploitation of OTEC potentials. Technology to fight corrosion and bio-fouling, for longer life of installed offshore machines, is also of vital relevance. The law of the sea is silent on the rights and obligations derived from energy extraction from the high seas and on disputes arising from such extraction, for instance, by a grazer, within a country's EEZ, upstream from another nation's EEZ, and on the issue discharging its return water above the thermocline, thereby degrading the downstream resources. While the worldwide OTEC potential is expected to be around 10 million MW, the same around the Indian coast has been estimated to be 50,000 MW (i.e., 0.2%) only.

Around 3000 kin along the Indian coast down from Bombay on the west coast up to Visakhapatnam on the east coast, a temperature difference of 20°C throughout the year is a available. Thus India has more than 3 lakh sq. km of tropical waters in her EEZ where sufficient temperature gradient exists throughout the year. Thus while estimated global OTEC potential is approximately 10 times the total world electric power production today, in India, the OTEC potential is about 15% of our total installed power generating capacity, hence necessitating the indigenous technological development and rapid scientific transformation relevant to this field.

According to one estimate by Kurachi, approximately 1 million tons of cold water must be pumped up per hour to produce 100,000 KW of electricity from OTEC plant 13, i.e. power needed for pumping operations alone is as much as 10% of the generated power. Even then, OTEC seems to be a useful addition to energy arsenal of the present world, if anyhow brought to the level of larger cost effectiveness. But it is too early in the development process to make any kind of economic assessment of the potential role of OTEC plants in the national and international energy picture. For what they are worth, cost estimates for very small plants seem very high even at experimental level. But what this means in terms of a possible full 400-500 14 megawatt plant, is simply impossible to determine at the moment. The history of technology transfer from government to prime industry and from pilot plant to operating levels is not one that would generate optimism. The USA with ultimate plan of 400 MW plants at and around isolated islands near Hawaii, Japan with 10 MW floating plants and 20 MW land based plants on the island of Nauru, France with 5 MW land based plant on the island of Tahiti, the Netherlands for islands of Ball, Sweden with proposals for the islands of Jamaica, UK with 10 MW floating plants and with plans of commercially profitable 40 MW plants, have either started functioning or are scheduled for the same in near future. The sums of money are very large, may be around US$60 million for a 10 MW floating plant. However in the Indian situation, as elsewhere, this baseload OTEC resource, by both of open and hybrid close cyles, seems to be of practical value favourably to southward situated islands where modest contribution to small and medium scale intraregional energy needs could be met successfully.

(E) *Wave energy:* Global wave energy potential is of the order of 45×10^{18} Watt. Variation of annual average value can be, for example, from 3 KW/m of coastline in the Mediterranean to 90 KW/m in the North Atlantic. The seasonal variation is remarkable i.e. in winter 6 times greater than the summer one. The principal recent studies have been made in UK, Japan, but interest has also been shown by France, Norway, Sweden, India, and the USA.Inventors have proposed many devices for large-scale wave energy collection and transmission to land, including, oscillating water columns, floats, moving pistons, rockers or paddles, and deformable bodies of both mechanical and pneumatic types.

However on a small scale, wave energy already provides the modest power needs of remote navigation buoys, including a large one, Kamei, in the Sea of Japan. In general, a 100m dome with 11 in diameter and 20 in high dome could generate 4000 KW from 2 meter high waves, a system known as dam-atoll system, made firstly by Lockheed Corporation. In simultaneously acts as a breakwater also which in turn, would have effects on the littoral movement and disposition of sediments. Thus, apart from small impediment to navigation and commercial fishing, the major problem of wave-power recovery has been from the very hostile marine environments viz. strong storms and winds. Besides, it may also be feasible to combine power dissipation with power extraction in future breakwater construction.

The use of wave power by navigation buoys will continue and should expand because it is a good example of on-site energy use, particularly at islands. However, at this primary stage of scientific achievement in this area, estimates of capital costs are highly speculative and subject to greater uncertainty, particularly when taking into account the likely future increase in the costs of energy derived from nonrenewable and depleting fossil fuels. In India, available informations based on visual observations indicate that annual average wave height along the Indian coast is of the order of 1.5 in with the wave period of about 6 seconds. But scientific data collected off Kalpakkam and near Kakinada on east coast, using wave rider buoys, indicate wave heights of more than 2.5 to 3 m, even during non-monsoon period.

This suggests that the annual average wave power potential along the Indian coast may lie anywhere between 5 KW/m and 15 KW/m. For total Indian coast, the wave energy potential is around 6,000 MW.Multipurpose Wave Regulatory Systems (WRS) are being employed and fixed near the appropriate sites off the Indian coast, so that further designing of wave energy devices could be done after taking into account the collection of climatic data, prospects of coastal protection from wave erosion and about other uses of coastal waters viz. aquaculture, transport, harbor developments, etc. Indian wave regulatory systems have been proposed to be set off the Madras coast at approximate water depth of about 10 meters and a distance of about 500 meters from the shore. Thus in India, a wave energy system exclusively for generation of electricity, does not seem to be economic.

A multifunctional system is being employed after incorporating the modifications to the oscillating water column reported in 1984 in Norway (i.e. The Harbour Concept). One twin turbine motor of 150 mm diameter has been successfully fabricated for demonstration, besides completion of studies on variation of wave energy at different depths, thus finding the optimum water depth, wave refraction studies in different lengths, completion of demonstration model on wave energy including a wave maker for a 1.2m wide flume and an oscillating water column to absorb the waves with a matching turbine proven in the 4 m wave flumes and generator to produce electricity.

(F) *Tidal energy.* Today France, the USSR, the UK and China are in the forefront of tidal energy development while the USA, Canada, the Republic of Korea, India and Argentina having made serious efforts towards its advancement, both in single and multiple type basins as well as single tide or double tide type. The peak efficiency of tidal power plants can reach 80%-90%, but for the simplest systems, the baseload is nil. Of the tidal power schemes now operating, only one is commercial in scale since 1967 with an output of 240 MW at LaRance in

France (with coast approximately US $120 million and produces electricity at a cost of 3.5 cents/kWh. The second one is an experimental one at Kislaya Cuba on Russia's Kola Peninsula in Barents Sea with some 400 kW capacity. In China large number of small plants in function is reported, while two sites of extreme potential are being developed by Canada and UK in Bay of Fundy and in Severn Estuary respectively, the later with estimated supply of 10% of the total England's annual demand for electricity. An alternative location across the Bristol Channel, west of Minehead, would even yield twice this amount. A recent review of the economics of the proposed tidal energy schemes in the bay of Fundy concluded that they were equivalent to the nuclear plants and markedly superior to the coal fired thermal plants.

The total tidal resource of world oceans has been estimated at 3 x 101 megawatts. Only 2% of these, at relatively 30 sites around the world, is suitable for harnessing at economically feasible level. Even then, it is significant because it represents 5% of the 1980 worldwide power generation from all sources. In other words, it is estimated that tidal power could displace between 5% and 10% of other energy sources in the area. Further, the plants have high reliability and a long life of around 75- 100 years and appear to increase four times in numbers in coming 50 years.India's Gulf to Kutch is one of the most suitable location.

Work has been in full swing to complete the festibility report by March 1987. The other possible sites, according to 1966 survey, are the Gulf of Cambay and the Sunderbans, having annual average tidal range of 6 m and 3.5 m respectively. The maximum tidal range found, yet in India, is in the Gujarat offshore areas i.e. around 11 m. According to Prof. E.M. Wilson, installed tidal capacities in the Gulf of Cambay, the Gulf of Kutch, and the Sunderbans are about 7300 MW, 100 MW, and 15 MW with estimated cost of around Rs. 1925 crores, Rs. 600 crores, and Rs. 15 crores respectively. The *7300* MW plant at Gulf of Cambay, if installed, will be able to contribute as much as 20% of India's total installed power capacity today.

Realizing the potential, Gujarat State Electricity Board and Department of Energy, Government of India have jointly taken up a detailed project study in collaboration with Electricity de France, along with 9 other interested government agencies. The project envisages construction of three main dams over the Hansthal Creek, the Sara Creek and the Phang Creek, having 3.1, 0.7 and 0.5 km. of length respectively.

The three dams will be connected with each other by a 30 km. long dyke carrying a road on its top, however the only proposed powerhouse will be over the Hansthal Creek, functioning for 12 hours a day, 6 hours in each tidal cycle, producing 600 MW of electricity. The CWPRS has already prepared a physical model of the tidal power project at Khadagwasla with a revised estimate of 5.3 crore, controlled by Kandla Port Trust. If the present siltation problem across here is solved anyhow, then this project will at least save Rs. 2-4 crores per year.

(G) *Currents:* Important currents, prevailing mostly in western parts of world oceans, include the Gulf Stream and the Guinea, the Brazil, the Mozambique, Japan (the Kuroshio) and the Australian currents. Both linear and rotary systems are being employed with large ocean coriollis turbines to derive this very diffuse energy source. For example, the Gulf Stream has the highest velocity (more than 2 m/s) and the power of its core with an average width of 50 km and depth of 120 in has been estimated at 24 GW. Assuming utilization of 20% at an efficiency of 40%, the resulting 2 GW of electric power would not amount to a significant fraction of world's energy supply. Besides the efficiency amounts to little more than 50%. However, if suitable submerged generators could be designed, they might be placed in gaps between atolls, where currents are often violent. Particularly if they are relatively close to the shore, these might be able to satisfy the limited energy needs of small neighbouring populations. Further, the navigational use of an ocean current can significantly reduce the transit time of ships travelling in its direction, which is a useful way of capitalizing on this energy source. The prospects of current energy on western coast of India is quite considerable, although not much development or technological breakthrough has yet taken in this field.

(H) *Offshore Wind Energy:* Placing wind turbines in offshore areas is environmentally advantageous. It reduces frictional losses. Although the net profit is only about 22%. Prospective site selection for installation of these turbines in India's EEZ, after surveying the different physical parameters in detail, is yet to be finalized. However, an imported wind generator has been installed for the first time in India to operate a lighthouse at Kanai Creek, Gujarat, with a power output of around *300* watts. The UK~ the USA, Denmark and Sweden are leading in this field of utilization of wind energy. Offshore wind generator systems would have to be relatively large in order to defray the high cost of power connection to the shore. In real terms, a successful wind power system is the sail driven vessel. If only a small fraction of the world's commercial transport fleet could be replaced by wind assisted ships, a significant amount of oil could be saved, seemingly of much importance for Indian foreign exchange savings.

(I) *Salinity Gradient:* Extraction of 240 mb of existing equivalent pressure by use of semi-permeable membranes on a very large scale has the potential to provide an overall substantial amount of energy. Its development might capitalize firstly on its abundance and secondly on possible low efficiency small plant. Further the power density of salt domes and brines, which is much larger than normal seawater, might make it a favoured application for salinity gradient density. World's major salinity gradient resource areas include coastal regions just offshore of major river basins such as Yangtze, Amazon, Congo, the Ganges an Mississippi rivers. In India, the Sunderbans and other large deltaic and estuarine points are potential places for generation of energy from salinity gradient.

Further in India, the total freshwater discharge from rivers an streams flowing into the sea amount to about 23 x 101 in sec., which could provide a power potential of about 54.8 101 MW. The energy density of this source is equivalent to about 250 in waterhead for a salinity difference of 35 ppt. However only a little work has been done on the technique of extraction of power from salinity gradient.

Other Sources

Although still at an experimental level, countries with larger EEZ have indeed possibility of extracting the following resources.

(J) *Magnetic ocean energy* production by magneto hydrodynamic generator (MHD) uses geotrophic currents and long strip that is wound with a flatwire, thus the flowing current produces electric current of 5 mho/m.

(K) *Marine biomass energy* in the form of CH^4 and CO^2 can be obtained mainly from Macrocystic and Glacilaria species. Current global estimates claim that marine biomass power could be competitive with conventional energy production. The prices would run about $18 per 100 m^3 with ayield of 112 tons/hectare/year. Major technological breakthrough is still awaited.

(L) *Marine chemosynthetic energy is* mostly associated with the vents near the seafloor great rifts, and from unusual marine faunas like worms, mussels, and clams found at depths around 2550 in and below using economically worthless H^2S for the purpose would be cheap and will also reduce pollution and eliminate a wate product, helping indeed in alleviating cost of the source material.

(M) *Geothermal energy* associated with rift zones and areas of intense volcanic activity are possible marine economic sources, particularly those who are coastal and insular in nature lying within the EEZ limit.

(N) *Heavy water viz.* deuterium and tritium, abundantly dissolved in oceanic water when fused at great temperatures releases great amount of heat, easily usable to generate electricity. Japan is seriously doing the pioneering work in its EEZ related to heavy water energy.

(O) *Radioactive energy* from marine sources is being developed. Japan Metal Mining Agency has started constructing experimental factories for the retrieval of uranium dissolved in seawater. Currently the Siemens Corporation of the USA is also attempting to use marine redioactive chemicals and radioactive wastes being dumped in oceans as energy sources. Current models are producing around 1.50 watts at a cost of $5 a watt and needs attention for possible economic viability after 5 years.

In a nutshell, we may comment in the field of other marine energy resources discussed just above, economic viability through capital investment and technological attainment, are the main points to be stressed upon in near future for any commercial level exploitation of these infinite renewable energy sources present in the ocean. The Indian situation in all these areas are yet at experimental exploratory level. Production, however, may be feasible within 15 years from now.

From the available geological and geophysical data, it has been estimated that the continental shelf around India up to a depth of 200 in, with an area of around 452,000 km^2, can have a potential oil reserve of 1 billion tonnes and natural gas around 271 bcm approximately. Recent offshore hydrocarbon discoveries indicate that the potential oil and natural gas reserve is very promising and we can look towards this in the coming 21st century with considerable optimism.

Most of the contemporary drilling activities in offshore area is carried out at depths of 200 in or so generally. Further, the Indian offshore oil reserves are of period Mioceane, Oligocene and Eocene. Proven offshore reserves represent a quarter of the world total reserves. About 90% of these reserves are estimated to be under the continental shelves, mostly under the respective coastal states jurisdiction. The overall strategy for the commercial development of offshore hydrocarbons depends on a careful appraisal of the potential for oil and gas and on technological and economic factors. Further much of the world's offshore oil and gas, mostly under EEZ, areas will be discovered and produced only if the relevant governments adopt a regulatory or fiscal policy appropriate to the realities involved.

Even the present reserve is considered to be maximum of 345×10^9 MT, feasible to be produced, it would not sustain consumption at current levels beyond the year 20 10. New findings of sufficient magnitude, able to sustain demand beyond the year 2000, would have to be made at twice the present discovery rate, but it can be visualized that even this optimistic scenario could only maintain A-situation in which oil and gas could be consumed as fast as it could be produced. In India most of the known petroleum resources are confined to the narrow belt of tertiary strata which constitute the outer margin of the extra-peninsular mountains along their whole line of contact with the peninsular block of Deccan viz. the Rann of Kutch, the Cambay Gulf off Piram island (dome), Bombay High and the Andaman and Nicobar Islands. The number of wells drilled in the offshore areas, by ONGC, was 36 in 1980 with a hole depth of 77,167 in against a total of 109 wells and 193,734 m of depth 32. According to the 1983-84 estimates, the ratio between production and consumption of oil in India is 1:1.3 approximately.

New discoveries have added to the attainment of self-sufficiency. Reduction in production as well as in consumption and efforts to supplement the fossil fuels by nonconventional and renewable sources of energy can only take us well within the 21st century, probably up to 2020 to 2025 A.D. Impressive achievement in last 15 years can be visualized with the fact that, in 1970, ONGC started offshore drilling, firstly near Aliabet in Gulf of Cambay. In early 1974, the Sagar Samrat drilling platform at Bombay High, acquired from Japan, started functioning, reaching production of more than 230.95 million tonnes in 1980, thus fully realizing the contribution of offshore oil reserve in national economy. In the field of natural gas, world proven reserve reached 90,325 billion cubic metres as per January 1, 1984.

World commercial production of gas in 1983 reached 1584 bcm. In India, the total reserve of natural gas according to 1980 estimates was 351.91 bcm, the offshore share being 270,96 bcm. 1982 total production was around 24.12 bcm with respective states contributions viz. Gujarat (48%), Assam (31%) and Bombay (21%). The 1980 and 1981 productions were respectively 19.25 and 19.97 bem. Location wise, natural gas (chiefly marsh gas with some other gaseous hydrocarbons) usually accompanies the petroleum accumulations, thus supplying the propulsive force which carries the oil from the oil-sands into the wells and if the pressure is sufficient, brings the oil up to the surface.

Since gas is both essential for the production of the oil and is also valuable as source of fuel on the oilfields, care is being taken now to prevent the huge waste of gas which was formerly so common. The country is fortunate to have discovered natural gas offshore the Andaman and Nicobar Islands, west coast of the Gulf of Cambay and Kutch and in basins of Brahmaputra, Godawari and Cauvery Basin. Principal present offshore supply being the Bombay High, producing 270.96 bcm. But Gujarat coast was on top in production of natural gas utilized between 1976-1980 i.e. 6.95 bcm, the others in series being Assam and Bombay High with value 4.55 bcm and 3.12 bcm respectively, making a total Indian production of 14.62 bem. However, the national objective, to be satisfied, is in need of compact and accurate programs with more developments in seismic surveys and interpretation capability, development of suitable shallow water indigenous mobile drilling platforms, much advancement in capability for design, fabrication and installation of fixed production platform, the same in the case of submarine pipelines, marine equipments, instruments, and sophisticated research vessels for further study, research, and resource development.

Seafloor deposits, within the EEZ jurisdiction of India, fit for present commercial production, comprise only of sands, gravels, corals, lime, shells and relatively small quantities of tin, titanium and iron. However potential seafloor mineral resources are, immense and includes both oozes and clays, phosphorites and manganese nodules, the rich muds and brines (found in oceanic rift areas) and the newly discovered polymetallic sulphides and cobalt crusts (in relatively deeper oceans). Sulphur is also being exploited but on a minor scale. Continental shelf rocks contain a variety of hard minerals, from coal and iron to tin, but only small quantities are being extracted commercially. For all this, ocean mining has to cope with availability, accessibility, advisability, and legality, among other factors.

Although Asia remains the continent with overall greatest promise of deposits of these valuable minerals, the superficial mineral deposits of Indian Ocean includes a wide variety of terrigenous, biogenous and authigenic mineral deposits. The beach and offshore place deposits of the Indian Ocean may be some of the largest in the world. The biogenous deposits comprise of the corals on shallow banks and on the continental shelves and the oozes in the deep sea. The authigenic deposits consists of the phosphorites, found both along continental margins (S. Africa and western India), and around seamounts in eastern and western Indian Ocean.

The continental margins of S. Africa, east Africa, southern Arabia, western India and the Andamans are marked by strong upwelling and provide nondepositional environments which are conducive to the formation of phosphorites. The polymetallic nodules found mostly in Central Indian Ocean Basin, beyond the national EEZ limit, cover an area of 10-15 $\times$ 106 km^2 and the resources are estimated to be about 1.5 $\times$ 10^{11} tonnes. In recent past, the exploration and exploitation of mineral resources of India's EEZ have attracted considerable attention.

Until the 1960's, the Indian Ocean was the least studied of the world oceans. It was mainly the International Indian Ocean Expedition of 1960-65, organized by UNESCO and ICSU, which considerably enhanced our knowledge of the ocean and mineral resources of this hitherto unexplored region. Recent exploration of mineral resources of lesser value, in shallow offshore areas, has actively started, seeing its growing importance. The prospects of offshore mining in India's EEZ were studied in detail in 1977. By now, survey of her total EEZ is supposed to be complete with maximum required oceanographic informations, later helpful in exploitation activity.

Tertigenous deposits: Occurrences of heavy mineral concentrates usually known as "black sands" are reported from many localities of Indian coast, which contain ilmenite, rutile, zircon, magnetite, monazite, garnet, kyanite, and tin in significant proportions. The offshore extensions of some of these placer deposits have been explored. But only offshore tin placers are being exploited at present. Most of the terrigenous deposits occur along the coast of Kerala between Quilon and Cape Comorin. The deposit is reported to contain 17 m tonnes of ilmenite, 1 m tonnes of rutile, 1.2 m tonnes of zircon, and 0.2 m monazite tonnes of. Ratnagiri coast (Maharashtra), Tinnaveli, Ramnad, Tanjore (Tamil Nadu), Visakha-patnam, Yarada, Waltair, Bhimunipatnam (Andhra Pradesh) and the coastal areas off Orissa.

Although the 25 km stretch along the coast from Neendakar to kayamkulam of Kerala state is supposed to have the richest deposit of offshore heavy minerals. Ilmenite concentration ranges from 17 to 79% in the area of Ratnagiri coast, forming an estimated reserve of 4×10^6 tonnes. Further exploration of offshore placers will be guided by the proximity to onshore placers and examination of the relict sands on the middle and outer shelves. The present data indicate that along much of the western shelf, from the Gulf of Kutch i.e. Vengurla to the vicinity of Mangalore, the relict sands consist of oolites and biogenous materials.

The extensive zone of potentially rich sands continues east to the Gulf of Mannar. However, it needs careful examination before starting the extensive exploitation. Worldwide offshore mining of sand and gravels currently exceed the offshore productions of all nonfuel minerals in both value and volume. But in the Indian case it has to compete with onshore deposits, mostly near coastal urban centres viz. Cochin, Madras, Visakhapatnam and Calcutta. But at the same time, the growing deepening of the sea near the land, is likely to increase coastal erosion.

Biogenous deposits : Besides shell deposits of Crassostrea madrasensis, C. cucullata, Villorits cyprenoides, Modiolus sps. and Metritrix sps., the other major biogenous deposits are the corals and the deep see pelagic oozes.

The best known deposits are in the Gulf of Kutch and in the Vambanad lake (back waters) of Kerala. Production from both these days being more than 1 million tonnes. Lagoons of the Laccadive Islands, the Andaman Islands shelf area and in the Gulf of Mannar and the Palk Strait appear to be promising. In general, an examination of the bottom sediment map within EEZ of the seas bordering India shows relict oolite and biogenous sands on the middle and outer shelves in the western coast continuously from the Gulf of Kutch to the vicinity of Mangalore and in patches along the eastern coast. Biogenous residues and the calcareous sands, amounting to 1135 m tonnes around Laccadive Islands can be economically attractive when exploited.

Regarding ornamental and semiprecious stones, India is becoming increasingly important as a supplier of marine ornamental shells and the exports have risen from 20 tonnes in 1969 to 466 tonnes in 1979. Further, India is yet not reached exploitation up to the optimum level, thus the way is still open for more exploitation and exports.

Authigenic (Chemogenous) deposits: The chemogenous deposits, formed due to rapid chemical reactions on the ocean floor in the vicinity of the Indian coasts, are found in the form of barium nodules off the west Andaman and the Laccadive Island coasts, phosphorite nodules off the north Andaman, and polymetallic nodules in the Central Indian Basin, of course beyond EEZ area bur under exploitable jurisdiction because of India being a pioneer investor.

In the area of production and reserves of phosphate rock, India is holding 13th position in the world, with her 1980 production measuring 0.6 million tonnes out of an estimated reserve of 108 million tonnes. This phosphate deposit is marked with vital presence of valuable uranium, 100 to 200 parts per million.

Algal and oolitic limestones, are also present off Saurashtra through Goa to Kerala coast. Further India's agro-dominated economy needs phosphorous as major fertilizer, thus making offshore exploration for phosphate deposits unavoidable in near future. Gypsum crustals are found in the inner shelf off the Maharashtra coast between Vengurla and Bombay in the form of silty clay or clayey silt.

A partial survey of EEZ of India has established that the offshore extension of heavy mineral placers on the western shelf are much larger than those available on land territory of India and coastal areas of other neighbouring countries. Further large areas of the seabed in the India's EEZ, both deep and shallow, are covered by biogenic sediments, but their economic viability is still in doubt.

But for islands like the Andaman and the Nicobar and the Laccadive, whose economies are becoming increasingly dependent on tourism, restraint would have to be exercised on the ruthless extraction of available mineral resource deposits. Systematic exploration and exploitation of Indian coasts for the phosphorite deposit is going to be indeed important in near future for purpose of self-sufficiency in fertilizers.

Exploration of India's EEZ is nearing completion now. Much emphasis should be placed on analysis and interpretation for greater understanding of the nature of deposits and technology needed for its proper exploitation.

Besides exact location of valuable mining sites and detailed worldwide study of market position of these derived minerals, both in terms of physical availability and economic valuation, is must before undertaking any major capital intensive mining business in hand. In fisheries, India occupies the 8th position in the world in terms of total annual catch, however the amount is the least in the world when calculated per fisherman engaged in this activity.

Regarding the need, based on futuristic studies, the projected fish requirements of India's population by the year 2000 A.D. is estimated to be 11.4 million tonnes. Her EEZ is still half-heartedly utilized as proved by the fact that 50.6% of total annual catch of over 4.47 million tonnes from the Indian Ocean is harvested from the narrow coastal zone only, i.e., area lying between the shore and 28 meters depth range.

The zone lying between 50 meters and 200 meters depth line contains 38.3% of total marine potential, while only 11. 1% of the total potential is there in the depth beyond 200 meters, i.e., 11.6% and 79.4% of area of total EEZ respectively. In other words 9% of the inshore zone of EEZ produces more than 50% of the fisheries.

Further the fisheries potential of the Indian Ocean is estimated at 14.4 million tonnes, of which the demersal fishes account for 7.4 million tonnes, the shoaling pelagic fishes for 6.0 million tonnes, the large pelagic fishes for 0.7 million tonnes, and the crustaceans and squids for 0.25 million tonnes.

Prawns are found mostly within 50 meter depth area of EEZ, the only known exceptions being off the south western coast in depths of 275 to 700 meters. Similarly a large trawling ground of deep water lobsters, discovered beyond the 50meters depth, was located between Quilon and Cochin. Tuna, having a wide migratory range, has been traditionally harvested from India's EEZ around the Minicoy and Laccadive Islands, although at lower levels..

Approximately 60% to 70% of the high priced tuna species of the Indian Ocean were taken by three non-Indian Ocean States namely Japan, Taiwan and South Korea. In 1976 India caught approximately 7-10% of the tuna and tuna-like fishes in the Indian Ocean. Yellowfin and skipjack are the two major species of tuna found within and close to the Indian EEZ.

Being the only underutilized tuna variety, skipjack, is in commercially exploitable potential in the Indian Ocean. India's catch of skipjack from Minicoy and Andamans through live bait method has gone up since 1973.

Besides tuna fishing requires highly specialized and technologically sophisticated operations in which India has little experience. Joint ventures and increased transfer of technology in the field of tuna fisheries is the need of the time.

Production through aquaculture has been increasing rapidly in recent past which can be evidenced with the fact that in 1980 its production was estimated to be 12.1% of the world's nominal catch, while in 1975 it was only 8.6%. India also leads in fin fish production in the world with an annual production reaching approximately 830,201 tonnes in 1981. But the fact remains that still 65% of the total Indian marine fisheries comes through traditional methods of fisheries only.

A comparative study shows that marine fisheries production in India has increased from 0.55 million metric tonnes in 1951 to over 1.6 million metric tonnes in 1980. On the economic front it contributes only about 0.5% of the GNP and 2% of the total export earning per year which, while considering its vast EEZ area, seems to be a very meager amount.

Of the total annual catch of around 2.5 million metric tonnes in 1980, 1.6 million metric tonnes were marine products; the aquacultures have been 850,000 metric tonnes, i.e., around 50% of the total marine products. However, by the year 2000 A.D. India could be producing 7 million metric tonnes through aquaculture. In the case of field prawn production in 1982, about 30,000 hectares of brackish water have been demarcated.

Under government sponsored plan it is proposed that at least around 5,000 hectares should be added annually for prawn culture in marine water. Thus in the field of mariculture an estimated 1.7 million hectares of estuarine or brackish water areas and 8.9 million hectares of inshore waters are to be utilized along the Indian coast within its EEZ limit.

For the successful resource management in Indian EEZ, the Indian government has recognized the main thrust areas for the optimization of investment at present, mainly high priority areas, manpower development has been sought at higher skilled levels which would successfully result in improving socio-economic condition of the vast population living mainly on the sea cost of the country.

Steps to start political and legal shells to understand the application of various process of UNCLOS III, mandatory transfer of technology and other associated complex issues like conservation, boundarly demarcation, pollution and protection of nation's interests in the EEZ are being sincerely undertaken. Comparative cost analysis in exploitation of these living and non-living resources is being seriously studied and their strategic importance being realized.

4 Global Warming on Sea Level

Geologists generally recognize that during ice ages the glaciation of substantial portions of the Northern Hemisphere has removed enough water from the oceans to lower the sea level one hundred metres below the present level, i.e., during the last and previous ice ages. Although the ice]covering much of the Northern Hemisphere have retreated, the world's remaining ice covers contain enough water to raise the sea level by over 75 metres. Estimates show that existing alpine glaciers contain enough water to raise the sea level by 30 or 60 centimeters. The Greenland and the West Antarctic ice sheets, each contain enough water to raise the sea level by about 7 metres, while East Antarctica has enough ice to raise the sea level by over 60 metres. However there is no evidence that either the Greenland or the East Antarctic ice sheets have completely disintegrated in the last two million years.

However, it is generally recognized that sea level was about seven metres higher than today during the last interglacial period, which was one to two degree centigrades warmer. Tidal gauges have been available to measure the change in sea level at particular locations over the last century. It was found that the rate of· sea level rise for the last fifty years would be between 2.0 and 2.5 mm/yr, while in the previous fifty years there was little change. Several researchers have attempted to explain the source of current trends in sea level rise. Barnett and Gornitz, Labedeff and Hansen, estimate that thermal expansion of the upper layers of the oceans resulting from the observed global warming of 0.4°C isotherm had migrated 100 metres downward, and concluded that the resulting expansion of ocean water could be responsible for some or all of the observed rise in relative sea level.

Meier estimates, that the retreat of alpine glaciers and small ice caps could be responsible for a current contribution to a sea level of between 0.2 and 0.72 mm per year. Concern about a substantial rise in sea level as a result of the projected global warming stemmed originally from Mercer, who suggested that the Ross and Filchner-Ronne ice shelves might disintegrate, causing a deglaciation of the West Antarctic ice sheet and a resulting six to seven metres, rise in the sea level in possibly 40 years. Subsequent investigations have concluded that such a rapid rise is unlikely. Hughes estimated that such a disintegration would take at least 200 years, and Bentley estimated that it would take even more than 500 years.

Hoffman, Wells and Titus estimated in 1985, that a warming of between 1° and 2.6°C could result in a thermal expansion responsible for contribution to sea level of between 12 and 26cm by 2050 respectively. In 1984, Meier estimated that a 2.8 mm rise had resulted from warming of 0.5°C and concluded that a 1.5°C to 4.5°C warming would result in a rise of 8-25 cm in the next century. Available estimates of the Greenland contribution assume that all meltwater flows into the oceans and that the ice dynamics of the glaciers do not change.

The NAS Polar Board in 1985 suggested that some of the water would refreeze decreasing the contribution to sea level rise, although a change in the dynamics might imply additional deglaciation and increase the rate of sea level rise. The panel assumed that such changes were unlikely to occur in the next century.

The potential impact of global warming on Antarctica in the next century, is the least certain of all the factors by which a global warming might contribute to sea level rise. Meltwater from East Antarctica might make a significant contribution by the year 2100. Although most attention has focused on projections of global sea level, impacts on particular areas would depend on local relative sea level.

Tidal gauge measurements suggest that relative sea level has risen 10 to 20 cm per century, more rapidly than the worldwide average along much of the U-S-coast. Local subsidence and emergence are caused by a variety of factors. Rebound from the retreat of glaciers after the last ice age has resulted in the emergence of Alaska and parts of Scandinavia. Groundwater pumping has caused rapid subsidence around Houston, Texas, Taipai, Taiwan, and Bangkok, Thailand, among other areas. River deltas and other newly created land subside as the unconsolidated materials compact.

Although subsidence and emergence trends may change in the future, particularly where anthropogenic causes are curtailed, no one has linked these causes to the future climate change in the next century. However, the removal of ice from Greenland and Antarctica would deform the ocean floors. Other influences on local sea level that might occur as a result of a global warming include currents, winds, and freshwater flow into estuaries. None of these impacts, however; has been estimated. A rise in sea level of 1 or 2 metre would permanently inundate wetlands and lowlands, accelerate coastal erosion, exacerbate coastal flooding, threaten coastal structures, and increase the salinity of estuaries and aquifers.The most direct impact of a rise in sea level is the inundation of areas that had been just above the water level before the sea rose. Coastal wetlands are generally found at elevations below the highest tide of the year and above mean sea level. Estimating past sea level rise has been the analysis of marsh peats commonly. Thus the impacts of sea level rise on wetlands are fairy well understood.

For the rates of sea level rise of the last several thousand years, marshes have generally kept pace with sea level through sedimentation and peat formation. As sea level rose, new wetlands formed inland while the seaward boundary was maintained. Because the wetland area has expanded one would expect a concave marsh profile, that there is more marsh area than the area found immediately above the marsh.

Thus, if sea level rose more rapidly than the marsh's ability to keep pace, there would be a net loss of wetlands. Moreover, a complete loss might occur if protection of developed area prevented the inland formation of new wetlands. Throughout the world, people have dammed, leveed and channelized major rivers, curtailing the amount of sediment that reaches river deltas. Even at present rate of sea level rise, substantial amount of land is converting to open water in Egypt and Mexico. Other deltas, such as the Ganges in Bangladesh and India are currently expanding seaward. These areas would require increased sediment to keep pace with an accelerated rise in sea level.

Additional projects to divert the natural flow of river water would increase the vulnerability of these areas to a rise in sea level. Several options have been identified for reducing wetland loss due to sea level rise. Abandonment of developed area inland or existing wetlands could permit new wetlands to form inland. In some cases it might be possible to enhance the ability of wetlands to accrete vertically by spraying sediment on them or restoring the natural processes that would provide sediment to the wetlands. Although coastal wetlands are found at the lowest elevations, inundation of lowland could also be important in some areas, particularly if sea level rose at least by one metre. Unfortunately the conventional ten feet contours in used the mapping of most coastal areas has prevented a general assessment of land loss.

Analysing date from aerial photographs to assess elevation in the area around Charkeston concluded that 160 and 230 cm rises could result in 30 per cent and 46 per cent losses of the areas dry land respectively. Such rises would result in 9 per cent and 12 per cent losses of the land in the area of Galveaton and Texas city. As with wetlands' loss the responses to inundation fall broadly into the categories of retreat and holding back the sea. Levees are used extensively in the Netherlands and New Orleans to prevent the flooding of areas below sea level and could be similarly constructed around other major cities.

In sparsely developed areas, however, the cost of a levee might be greater than the value of the property being protected. Moreover even where levees prove to be cost effective, the environmental implications of replacing natural shorelines with man-made structures would need to be considered.Sea level rise can also result in the loss of land above sea level through erosion. In 1962 it was shown that the erosion resulting from a rise in sea level would depend upon the average slope of the entire beach profile extending from the dunes out to the point where the water is too deep for waves to have a significant impact on the bottom (generally a depth of about 10 metres.) By comparison, inundation depends only on the slope immediately above the original sea level. Because beach profiles are generally flatter than the portion of the beach just above sea level, if implies that the erosion from a rise in sea level is several times greater than the amount of land directly inundated. Processes other than sea level rise also contribute to the erosion process including storms, structures currents and along shore transport. For example, water levels along the great lakes can fluctuate over one metre in a decade.

Geologists basically believe that coastal barriers can maintain themselves in the face of the slowly rising sea level through the landward transport of sand which washes over the island during storms, building the island upward and landward. The potential erosion from a rise in sea level could be particularly important to recreation beach resorts, which include some of the world's most economically valuable and intensively used land. Relatively few of the most densely developed resorts have beaches wider than about 30 m at high tide.

Thus the rise in relative sea level of 30 centimeters projected in the next 40 to 50 years could erode most recreational beaches in developed areas, unless additional erosion response measures are taken. Therefore, in many undeveloped countries, relatively inexpensive houses are found very close to the shore. Because the value of these houses is less than the cost of protecting them, they must be moved as the shore erodes. An accelerated rise in sea level would speed this process of shoreline retreat. A rise in sea level could increase flooding and storm damages in coastal areas for three reasons: erosion caused by sea level rise would increase the vulnerability of communities, higher water levels would provide storm surges with a higher base to build upon, and higher water levels would decrease natural and artificial drainage.

In 1984, Leatherman conducted an analysis of Galveston Island, Texas. He estimated that the area within the 100 year flood plain would increase from 50 per cent to 94 per cent for an 88 cm rise in seal level and that for a rise greater than one metre. The Galveston seawall would be overtopped during a 100-year storm. A wide variety of shore protection measures would be available for communities to protect themselves from increased storm surge and wave damage due to sea level rise. Many of the measures used to address erosion and inundation including seawalls, breakwaters, levees and beach restoration also provide protection against storms. In the case of Galveston which is already protected on the ocean side by the seawall, it might be necessary to completely encircle the developed areas with a level to prevent flooding from the bay side, upgrading the existing seawall might also be necessary.

Although most researchers and general public have focussed on the increased flooding and shoreline retreat associated with a rise in sea level, the inland penetration of salt water could be important in some areas. Besides as a rise in sea level increases the salinity of an estuary represents the outcome that (1) the tendency for the ocean salt water to completely mix with the estuarine water, and (2) the tendency of fresh water flowing into the estuary to dilute the saline water and push it back toward the ocean.

During the rainy season, therefore, low salinity level prevails. A rise in sea level has an impact similar to decreasing the freshwater inflow. By widening and deepening the estuary, sea level rise increases the ability of saltwater to penetrate upstream. Hull, Thatcher and Tortoriello in 1986 examined the potential impacts of an accelerated rise in sea level due to the greenhouse warming. They estimated that 73 ern and 250 cm rises would result in ocean front migrating as much as additional 15 and 40 kilometres, respectively. Thus the causes and effects of sea levels rise are a matter of grave concern and it, is now understood that if proper steps to restrict the rise are not taken, then dangerous consequences would result in a magnified way and affect the human life in a very horrific manner, because it is to be clearly understood that human life would also be affected by the rising sea level, especially of the people in the coastal areas.

Thus it can be concluded that the problems of future sea level rise are important not only to geomorphologists, ecologists, geographers and social scientists, but also to governments and their agencies concerned with planning and management of the coastal regions that will be modified by the changes. Of late, it has become increasingly clear that man has acquired the power and capability to alter the ecology of this planet adversely. Researchers in the past few decades have clearly demonstrated that urgent attention is needed to reverse this trend. The influence of the oceans which cover nearly 71 per cent of the earth's surface, is of much importance for regulating the climate of this planet. The last decade has witnessed sudden increase in man's awareness of the oceanic environment, and the progressive deterioration of our seas due to human activity.

Any change in the equilibrium of the sea, however minor it may be, is bound to affect the human population, as 20 per cent of the total global population is living in the coastal belt of the landmass. Therefore, it is a matter of utmost importance to develop an understanding of the ocean and coastal environment. Moreover, the changes in the sea level is a subject which deserves our immediate attention.

The global warming, an important factor for sea level rise, is caused by the release of carbon dioxide, chlorofluorocarbon, nitrous oxide and methane in the atmosphere due to human activity. The consequences of sea level rise are far reaching and these result in increased coastal erosion, submergence of coastal wetland, increased damage due to storm surges, shifting of the river mouths and deltas, salt water intrusion into the rivers, etc. In addition, ecological damages include loss of coastal vegetation, loss of commercial fisheries, destruction of important nursery grounds like the mangrovers and extinction of valuable coastal fauna and flora.The most severe impacts are likely to stem directly from inundation.

Asia would be most affected because of the extreme vulnerability of several large and heavily populated deltaic regions. For example, with a 1.5m sea-level rise, about 15 per cent of all and (and about one-fifth of all farmland) in Bangladesh would be inundated and a further 6 per cent would become more prone to frequent flooding. Altogether 12 percent of agricultural production could be lost. In Egypt, it is estimated that 17 per cent of national agricultural production and 20 per cent of all farmland, especially the most productive farmland, would be lost as a result of 1.5m sea-level rise. Island nations, particularly low-lying coral atolls, have the most to lose. The Maldive Islands in the Indian Ocean would have one-half of their land area inundated with a 2 inch rise in sea-level. In addition to direct farmland loss from inundation, it is likely that agriculture would experience increased costs from salt water intrusion into surface water ground water in coastal regions.

Deeper tidal penetration would increase the risk of flooding and rates of abstraction of ground water might need to be reduced to prevent recharge of aquifers with sea water. Further, indirect impacts would be likely as a result of the need to relocate both farming populations and production in other regions. In Bangladesh, for example, about one-fifth of the nation's population would be displaced as a result of the farmland loss estimated for a 1.5 m sea-level rise.

The coastal regions around the world, particularly areas near the mouth and along the great rivers have been centres of population concentrations. Trade, commerce and industry have flourished well because of easy access to the resources of the hinterland and reach of the markets across the oceans. Thus, it is not surprising that a significant portion of the world's population lives within the coastal zone with many of the cultural artifacts, built at elevations less than 3 m (10 ft.) above mean high-tide level along the shoreline. Even without a change in the height of mean sea level or the height of the land, the elevations of settlements and other artifacts are not adequate to provide safety in the event of extreme stormsurges, etc. The hazard of flooding, erosion, intrusion of saline water in the underground aquifers has grown increasingly apparent and serious along many of the world's coastline as local mean sea level has risen during the twentieth century at a rate ranging from 1 to 5 mm/yr.

The scientific and engineering community has focused their attention of the problems of accelerated beach erosion and the rising level of green house gases in the atmosphere. As a first approximation, the effects of one metre sea level rise on a coastline in South Asia may be determined by surveying the contour one metre above present high spring tideline, but allowance must then be made for the effects of erosion and deposition as submergence proceeds. Submergence will result in recession of the coastline, except where there is vertical cliff, or where deposition of sediment continues at a sufficient rate may continue in the civility of the mouths of major sediment-yielding rivers, such as the Ganges, which have built large deltas. Submergence are likely to accelerate the recession of cliffed coasts, except where the rock outer crops are resistant, and high and low tide lines simply move up the cliff face.

In the tropics cliffs are mainly on headlands, such as Cape Comorin in India and Condra Head in Sri Lanka, and the rock outcrops are usually deeply weathered and easily eroded. Many are vegetated bluffs rather than actively receding cliffs, and a sea level rise is likely to understand them, producing slumming and eventually steep, active cliffs. Cliffs tnat are already receding, as on the coast near Chittagong and Cox's Bazar in Bangladesh, will show an acceleration of cliff retreat as sea level rises, and larger waves move in through deepening water. Where shore platforms have developed, they will disappear beneath the rising sea. Some will be built up by nearshore deposition, or the growth of organisms such as algae and corals.

Submergence of low-lying rivers deltas and coastal plains will cause extensive retreat,, of the coastline, accompanied marine erosion, but a few sectors may be maintained by sedimentation, especially in the civinity of the mouths of large rivers. On the north coast of Java there is already erosion on deltaic sectors where former river mouths have been diverted. Naturally or artificially, but this is a largely compensated by the provocation that continues around the present river mouths. With a rising sea level, the proportion of eroding coastline will greatly increase on these and other deltas.

In India, Ahmed estimated that a quarter of the coastline was eroding as the result of delta growth, but this proportion will be much reduced. As sea level rises, beaches will become narrower, and beach erosion will become more extensive than it is now. An exception may be found locally where beaches are nourished with a larger supply of sediment derived from rivers draining areas of increasing rainfall, or from accelerating clifferosion, but some of the sediment supplied to the coast by rivers will be retained within submerging river mouth areas. As submergence proceeds, coastal inlets, embayments, and estuaries will enlarge and deepen and salinity penetration from the rising sea will increase. Tidal regimes will be modified, and intertidal shoals, mangroves and marsh islands (as at the mouth of the Ganges) will show' extensive erosion as these features disappear beneath the sea. Depositional terraces built up by mangroves and salt marshes during the Holocene slits and will only be maintained if sedimentation continues at a sufficient rate, but generally their seaward margins will be cut back. Coastal lagoons of the kind found in Southern India and around Sri Lanka sea will erode their enclosing barriers in some cases submergence and erosion of barriers will reopen lagoon and coastal inlets or embayments.

Fringing and nearshore coral reefs will be submerged in the name way as rocky shore platforms fronting cliffed coasts, but the rising sea level may stimulate the upward growth of corals and associated organisms, which could maintain the reef structure if they keep pace with the marine trans region. This will be critical for the future of low sandy islands (cays) which have formed on coral reefs flats as in the Maldives and the Andaman islands.

Low-lying coastal areas will show migrations of vegetation and faunal zones as submergence proceeds. Mangroves are extensive in such areas as the Sunderbans and the tidal shores of the Ganges delta, but mangrove fringed coastlines have become less extensive in Southeast Asia in recent decades because of the impacts of land reclamation, fishpond construction, mineral dredging, and waste disposal.

In Thailand, for example, these impacts have already reduced the mangrove area by more than one-third. Where mangroves persist, they typically stand in front of zones of salt marsh and freshwater vegetation (for scrub or swamp forest) where there has been coastal degradation, the advance of the mangroves (often arranged in species zones) has been followed by salt marsh and fresh water communities in succession as sedimentation builds up the substrata to appropriate levels, forming an intertidal depositional terrace.

A sea level rise will reverse this sequence, unless sedimentation continues at a sufficient rate to maintain or propagate the coastline. Submergence will kill the sewad mangroves, and initiate erosion of the previously prograded terrace.

As the coastline retreats, the mangrove zone will migrate landward to displace the salt marsh, which in turn will invade the freshwater hinterland. Mangroves regenerate quickly in areas that have been cleared, then abandoned, and it is likely that they will spread back on to suitable low-lying hinterlands as the sea rises, but where the hinterland is steep or rugged the mangrove zone will be extinguished by submergences.

In South and South-East Asia, the land immediately behind the mangroves is commonly used intensively for fish ponds or rice fields and these would be invalid by the retrogressing mangroves if local people permit. In drier areas, such as the Rann of Kutch where mangroves are backed by base, hypersaline plains, a sea level rise will result in the colonisation of presently unvegetated tracts by mangroves and the coastline retreats.

Coastal lagoons typically show salinity gradients from fresh river inflow through brackish to sea water at their marine entrances; this is true for example in the chain of lagoons that opens to the sea at Songhkla in South Eastern Thailand. A sea level rise of a metre over the next century is likely to increase marine inflow through deepened and sidened marine entrances. Where this occurs areas that were previously fresh water will become blackish with dieback of existing vegetation and replacement by mangroves, salt marsh and other salt tolerant hydrophytes. There will also be displacement of fresh water fauna by estuarine and marine species. In these various ways ecosystems in coastal regions will change as sea level rises on the coasts of South and South East Asia. The natural vegetation and fauna have already been extensively modified and what actually happens will depend very much on human responses to ecological change.

The way in which modern coastal societies will respond to be a relatively rapid rise in sea level will vary with political and economic factors. Nations that have the organization, technology and resources to counter the effects of a sea level rise by means of artificial structures will do; so others will, in general have to choose between evacuation of submerging areas or adaptation to changing coastal environments. On coastline close to urban and industrial centres it is likely that a sea level rise about a metre over the next century will stimulate expenditure on structures designed to prevent submerging and erosion.

It is very likely that where land has recently been reclaimed from the sea for use by large coastal population (as in Singapore), strenuous efforts will be made to retain it, by building sea walls and introducing pumping systems using the techniques familiar from the history of the Netherlands coasts. Where Erosion Beaches are valued for their creation or tourism, as in seaside resort such as Pattaya in Thailand it is possible that they will be maintained by beach refreshment programmes of the kind already used in the US and Australia.

More generally beach erosion will continue to be met by the spread of artificial structures; already more than 260 kms. of the Kerala coastline in South India are fringed by granite boulders walls.Away from highly developed centres there may have to be abandonment of coastal fringes as submergence proceeds and modification of developed land and water uses as the water table rises and soil and water salinity increase.

A sea level rise of a metre over the next century will accentuate such problems on the intensively utilised and densely populated coastal plains of South and South-West Asia, causing coastline recession of up to several kilometres, displacing coastal villages and depriving many people of their land and resources.

Tide range in right of Bangkok, is up to 2.3 metres and a sea level rise of a metre point gently sloping coastal plain would submerge the whole of the reclaimed coastal plain would submerge the whole of the reclaimed mangrove area and the additional zone up to five kilometres to landward existing fish ponds, shrimp ponds and salt pans are thus likely to be submerged and destroyed by marine erosion unless they are protected by even larger and more elaborate sea walls.

The predicted sea level rise of a metre during the coming century will greatly modify the geomorphology and generate a whole new range of problem for the communities that occupy and utilize these coastal areas. The effects of a sea level rise can be reduced theoretically or inferred from studies of coastlines where submergence is already in progress, but there is a need for site specific predictive studies of the physical and ecological consequence of a sea level rise on the various types of coastline, especially those that are low lying densely populated and intensively utilized. The human response to sea level may be counter measures, where resources and technology are capable of these but elsewhere there will have to be evaluation or some kind of adaptation to the changing coastal environment.

On the coastlines of South and South-East Asia, the most likely outcome is an intensification of use and occupancy of near shore sea areas.Global climate change when talked of brings about many such changes, which directly or indirectly affect human activities, the social, economic and political set-up.

This sea level size directly effects the life of human beings, especially those, who live in coastal areas. Their economics and politics are very much dependent on the sea, and any changes in the sea will naturally bring about changes in their life-styles, economics and politics, equally, as all three are interrelated.

Ecologically, sea level size, brings about consequences like the changing of the area's biosphere, which would naturally affect human activities, as the people living near the sea, would be very much dependent on the biotic resources, especially those obtained directly from the sea, like fish, and other such sea foods, pearls etc.

Submergence of coastal wetlands, inundation, coastal erosion, flooding and storm damage, especially in the coastal erosion, flooding and storm damage, especially in the coastal areas, increased salinity in estuaries and aquifers, and all these factors would affect the human life to such a degree that, special efforts are required to right these consequences, brought about by the increasing sea level. As a consequence, the coastal zone is to be regarded as one of the most important and critical parts of the planet, particularly as far as global change is concerned.

On the one hand, the physical context is transforming because of changing sea level and progressive coastal erosion; on the other hand human communities concentrating along the sea exploit its coastal resources. There is a growing awareness today that human activities are beginning to affect out plant on a global scale. Sea level rise and other potential consequences of climate change create problems that can be solved only by international cooperation.

There is an interaction between social conditions and the impacts of sea level rise. On the one hand, particularly if coastal defence is impossible there could be a large social impact in the form of the abandoning coastal areas and the migration of population whether or not this is considered acceptable, depends on the values and norms of the society.

On the other hand, the way in which a society responds to threats from a rising sea level depends very much on the social structure. Seventy per cent of the human population lives within 60 km of the sea, and the percentage is increasing. Human settlements, industry and agriculture in coastal lowland are vulnerable to rising sea level or changes in storminess or storm surges. Some nations are particularly vulnerable. Eight to ten million people live within one metre of high tide in each of the unprotected river deltas of Bangladesh, Egypt and Vietnam.

Half a million people live in archipelagos and coral atoll nations, that lie almost entirely within three metres, of sea levels, such as the Maldives, the Marshall Islands, Tuvalu, Kiribati and Tokelau. Other archipelagoes and islands nations in the Pacific and Indian oceans and Caribbean could lose much of their beaches and arable lands, which would cause severe economic and social disruption.

Even in nations that are not, on the whole, particularly vulnerable to sea level rise, some areas could be seriously threatened. Examples include Sydney, Shanghai, coastal Louisiana and other areas economically dependent on fisheries or sensitive to change in estuaries habitats.

As a result of present population growth and development, coastal areas worldwide are under increasing stress. In addition, increased exploitation of nonrenewable resources in degrading the functions and values of coastal zones in many parts of the world.

Consequently, populated coastal areas are becoming more and more vulnerable to sea level rise and could have serious adverse effects. A rapid rise of the mean sea level 1 to 2 metres in 100 years will have a considerable impact on all environmental conditions in low deltaic areas, coastal marshes and embayments in coastal lowlands. They are often densely populated, and highly productive in the agricultural sector. The impacts on morphology and water quality will lead to counter measures in the fields of flood protection and water management.

A common situation is that of a growing delta. The land areas are built up by sediments from the river and the sea, to an elevation slightly above high tide in the coastal strip and to the level of 1 to 5 years flood along the rivers. So the land level rises and the delta acerbates into the sea. The whole process is strongly related to the local relative change of mean sea level.

A rapid increase of the global rise of mean sea level will upset the present dynamic equilibrium. The process will adapt to the new rate of rise which generally means a slowing down of the rate of accretion even recession and drowning may be the result. A rise in mean sea level will also influence the propagation of the tides and the generation of storm surges.

Sea dikes will have to be high often even more than the rise of mean sea level. In some cases a shortening of the coastline by enclosure of embayments may be feasible. The rise of sea level and the related backwater effects increase the water levels in the estuary and, with some time lag, in the lower part of a river. In these areas also the floor protection has to be adapted.

From a geological point of view deltas belong to the most rapidly changing parts of the crust of the earth. A common situation in cases where man has not yet interfered or only to a limited extent is that of a growing delta. This is a delta which is extending into the sea entailing a rise of the riverbed and hence of the levels of the river floods. The land areas are built up by sediments from the river and from the sea. Under normal conditions the land in the costal strips build up to an elevation slightly above high tide level to approximately the level of the 1 to 5 years flood.

A rapid rise of the sea level within a certain period will upset the present dynamic equilibrium. There will be a gradual adoption of the morphology to the new sea level. Because of the time lag the water depths of the river channel and the sea floor will initially increase slowing down the rate of building up.

The tidal currents in the sandy coasts. The mangrove belt which exists along the disappear. This belt which exists along the coasts of most tropical deltas will deteriorate or even disappear. In deltas where the distribution of the upland discharge over the tributaries has not been stabilized by hydraulic works, a river branch may be cut off from the supply of sediments (morbund river). The part of the delta which was supplied with sediments by that branch will be exposed to erosion in the case of a rapid rise of the sea level (parts of the delta of the Mississippi River).

An extreme case is that of the drowning delta like the (former) delta of the Rhine and the Meuse in the Netherlands. Now supply of sediments, natural and man-induced subsidence and a rise of the sea level have led to an artificial situation where man is constantly fighting the destructive forces of the sea. Man has learned how to cope in this area with a continuing relative rise of the sea level which, if man induced subsidence is included, amounts to some 3 metres during the past 1000 years.

Coastal embankments are exposed to high water levels and, depending on the situation, to impact of waves. High sea levels are caused by astronomical spring tides and along a number of coasts, by wind-induced set-up of the sea level (storm surges, cyclones, typhoons and hurricanes).

A rise of the sea level will affect these conditions in different ways. The propagation of the astronomical tides in shallow coastal waters will be modified, increasing, in general, the tidal range. The set-up will decrease but, in general, this effect will be small. The wave uprush may substantially increase, especially in those cases where under the original conditions, the depths in front of sea dikes are small or where tidal foreland exists.

The result is that the sea dikes will have to be raised to a greater height than the rise of the sea level to maintain the same degree of safety. In many deltas embankments have been raised periodically to cope with the relative sea level rise. The motive has often been not so much for this rise but the desire to ensure a greater degree of safety, especially after disasters. This was also the motive for the implementation of schemes aiming at shortening the lines of defence by enclosures near the mouth of estuaries and other tidal embayments.

In all estuaries where a fresh water river debauches into a basin with saline water the heavier saline water will penetrate into the estuary and more in an upstream direction in spite of the flow of freshwater in the opposite direction.

Depending on various factors, in the first place the tidal range in the basin, the estuary will show a distinct interface between the fresh water near the surface and the saline water near the bottom or a situation will occur when there is a partial or complete mixing. The saline intrusion renders the water unsuitable for various purposes and is a maximum when the riverflow is minimum during dry periods when the need for fresh water is high.

An extreme case is the Gambia river in West Africa where at minimum flow the saline effect is felt up to 235 km. upstream from the sea. A rise of the-sea level will increase the intrusion of saline water because of the increase in water depths, at least in a first phase, and the increase in tidal flood volumes. The shift of the saline reach in the upstream direction will affect the intake of water from the river for irrigation and water supply and the ecological conditions of the channel and the adjacent land. It should be noted that this process is already taking place in a number of estuaries where because of dredging in connection with post development the saline effect has moved further upstream.

A classical example is the Rotterdam waterway giving access to the port of Rotterdam where the saline reach moved over 35 km. in a period of about 60 years.Low land rice is grown in many deltas in the humid tropical zone where storm surges occur and where there is no protection against river floods. In these areas, often there are only low coastal embankments to protect the coastal strip from flooding with saline water during high astronomical spring tides. Local rice varieties are grown which are adapted to the depth of flooding caused by river.

Near the coastal strip the depths of flooding are small and relatively high yields are obtained. Further upstream the depth of flooding maybe as much as 3-4m and floating rice' is grown with low yields. These conceptions occur in many parts of the large deltas of South East Asia and especially in the delta of the Mekong river. A rise of the sea level of, say 1 m or more, would upset these conditions, the sea-dikes will have to be raised and in the tidal reach the depth of flooding will increase affecting the productivity. To maintain the present productivity levels protection against flooding will become necessary and embankments will have to be created.

Although, in itself, this is a simple operation, it will change the environment profoundly and will entail many hydrological and morphological side effects (rise of the flood levels, rise of the river bed and increased meandering).

Embanking makes it necessary to lay out a system of drainage canals and outfalls. Because of the elimination of the beneficial effects of the floods (supplemental irrigation and flushing it may become necessary to provide irrigation and flushing) it may become necessary to provide irrigation facilities. Under the present conditions with the continuing relative sea level rise and the desire to increase the yields and to grow other crops than rice there is a tendency to carry out development schemes in the deltas aiming at flood protection, drainage and irrigation.

An uncertain economic feasibility often hampers actual implementation. A rise of sea level which far exceeds that occurring at the present will either ruin many agricultural areas or accelerate the developments which are already taking place. Depending on the land elevation, relative to tidal levels, and on the drainage requirements (low and rice or dry food crops) either gravity drainage or pump lift drainage can be applied. In the coastal zone of many deltas all over the world tidal drainages is applied on a large scale when the land is not too low. The system often consists of simple wooden box sluices with glap gates.

If there is a significant rise of the sea level this system will have to be replaced by pump-lift. This will entail high installation and operation in deltas in the humid tropical zone with their high amounts of rainfall. Where pump-lift drainage is applied like in the Netherlands, Italy, Japan, USA and India the pumps will operate at lower efficiency if there is a substantial rise of the sea level and the structure may have to be modified or redesigned. The sea level rise will increase the salt water intrusion threatening the intake points of irrigation water along the river. Unless the drastic solution of enclosure of the estuaries near the mouth is applied, the intake points will have to be shifted upstream beyond the salinity limit during periods of low river flow.

New canals will have to convey the irrigation water from the new locations of these intakes to the downstream areas and when the stops are small booster pumping stations in the canals will be necessary. A benefit of the sea level rise will be the increase in head for the flow of irrigation water from the river to the land. If there is a rapid rise of sea level cities in low-lying coastal areas will have to be protected from flooding by encircling embankments (pocket dikes) and in any case, will have to be drained by pump-lift. This is an operation which is already under way in a number of large cities mainly as a consequence of man-induced land subsidence.

Typical examples are New Orleans and Osaka, Bangkok is a recent case. In Osaka a (local) maximum subsidence of 2.8 m occurred in a period of about 30 years. This probably is a greater rate than any conceivable rate of rise of the sea level. The walls along the city canals and the entrances from the sea have been raised by 1 to 3 m. check-sluices (some with vizor gates with a span of 60 m) have been built which can be closed when high typhoon levels occur on the Pacific ocean. The city area is then isolated from the surrounding waters and excess water from local typhoon rainfall has to be removed by pumping.

The Kema pumping station, probably the largest of the world, with a capacity of 330 m/s (pumps of 55m/s) and a head of 3-4 m has been constructed for this purpose. The response of the Japanese to this challenge has been immediate and vigorous. Because of financial and organizational problems this however would probably not be possible for cities like Shanghai and Calcutta, should a rapid relative rise of the sea level occur which would be similar in effect to the subsidence of Osaka.The case of Osaka demonstrates that, from the technical point of view, area of the sea level of say 2 to 4m within a period of 100 years would not cause unsurmountable problems. The case has been justified because of the high economic value of an urban place like

Osaka. In tidal harbour basins it will be necessary to raise the heights of quay walls and landing stages. Attention should also be paid to ensure that there is sufficient headway under bridges. In the case of dock harbours the access locks may have to be adapted to a rise of sea level. In a number of coastal areas in France, India, Japan, the Netherlands and to a lesser extent Bangladesh and the Mekong-delta coastal or estuarine reservoirs have been framed by damming of tidal embayments or estuaries. Excess water frcm the rivers is discharged into the sea through sluices and the enclosed basins have been keen transformed into reservoirs with fresh water by the inflow of river water.

The discharge of to the sea is effectuated by tidal drainage. The low lying areas around the reservoirs can abstract water from these reservoirs during dry periods; generally they also drain to these reservoirs during rainy periods, either by gravity or by pump lift. A rise of the sea level of around 0.5 m or more will completely upset the present operation of these reservoirs, since gravity drainage into the sea will no longer be possible.

There are two basically different solutions which could be considered. The one could be, that the reservoir levels would be maintained at the original level. This implies that excess water from the reservoir would have to be removed by pumping and that the sluices would be closed. Since the volume of flood waters from the river may be quite high, this could be an expensive solution.

In the second solution operational level of the reservoir could be raised, so that gravity drainage to the sea remain possible. This would require modification or redesign of the existing sluices. The drainage of the surrounding low lying areas on the reservoir would be --hampered and pump-lift may be necessary. Embankments would have to be raised not only along the reservoir- but, because the rivers debauching into the back-water effects, also along the rivers debauching into the reservoir. Indeed in a delta, with its system of interconnected watercourses, the side of the reservoir level would propagate in the entire system unless partitions are made. There would also be seepage flow from the reservoir with its high level to the surrounding lowlying areas, but after a long period of time this seepage water would have the same composition of the water in the reservoir. A large population of the world lives in coastal areas and on islands. These areas have several assets, including some of the largest cities of the world, ports, industries, fertile lands, fisheries and tourism facilities.

A sea level rise of one metre by 2100 would have severe impact on all coastal areas particularly in the developing countries. More frequent recurrence of storm surges would pose an intolerable burden on the coastal population, impact on shorelines erosion, inundation of wet lands, and mangroves, increases in the salinity of aquifers, and drowning of coral reefs would cause severe ecological damage. The small islands are particularly vulnerable.

Human beings in their responses would naturally try to adapt to these changes and the rising sea level especially in the coastal areas where the effect will be such higher. This adaptation will be brought about, because of the severe stress on the economies of those effected areas. Agriculture will be affected as sea level area could inundate low lying coastal plains with fertile alluvial soils where a high proportion of rice production is located. Fisheries could be affected. Two-thirds of the world's fish catch and many marine species, depend on coastal wet lands for their survival. Without human interference eco-systems would migrate landward as sea level rises and then could remain largely intact, although the total area of wetlands would decline and so would the fish catch. If human intervention is followed then there are chances, that a large portion of these ecosystems would be lost which again would reduce or finish the fisheries. While it is very difficult to estimate the economic implications correctly, it to possible to make some broad statements about the various options. In densely populated and productive areas retreat may not be viable in general and in the case of small islands, could be impossible.

Accommodation could be extremely costly but could be justifiable in particular situation. Protection would be economically viable, in particular cases but could be unaffordable for poorer countries. One estimate quoted in the Inter-governmental Panel on Climate Change report gives the total additional costs to be more than 500 billion dollars with a major part to be incurred in developing countries. Thus a gentle coherence is needed and to understand the human response, they can be put into three categories.

(1) Retreat, i.e., abandon vulnerable areas and resettle-inhabitants.

(2) Accommodation—adjust to sea level rise

(3) Protection for vulnerable areas particularly those with high population, density, economic value or precious natural resources.

Retreat: Preventing development can reduce future expenditure foe adaptation. India, Sri Lanka, USA, Australia and a number of island nations have already such regulations. Where the area that could be immundated is high, the timing of such regulations would be difficult. Spreading of awareness among all the occupants of vulnerable areas could encourage private initiatives and solutions. This is no solution, however for small island states and poorer countries. Also under this option, resettlement could create major problems. Resettled people are not always well received. They often face language problems, racial and religious discrimination and difficulties in obtaining employment. Even when they feel welcome, the disruption of families, friendships and traditions can be stressful. Although the impacts of accommodation and protection would be less, they may still be important. The loss of traditional environments which normally sustain economics and cultures and provide recreational needs could disrupt family life and create social instability regardless of the response eventually chosen. Community participation in the decision making process is the best way to ensure that these implications are recognized.

Accommodation: This response requires advance planning. Many coastal structures could be elevated for protection from inundation. Building-codes could be devised for protection from surging waters and high winds. Drainage could be modified, storm warning and preparedness plans could be strengthened. Where salt water damages agricultural lands, salt-tolerant crops may be introduced. Some agricultural lands may be converted to agricultural uses. Human activities that harm coastal resources maybe banned, e.g. filling wetlands, damming rivers, mining coral and beach sands and cutting mangroves. Undeveloped land can be set apart to accommodate re-establishment of wetlands and mangroves.

Protection: This response involves defensive measures against in addition, tidal flooding, shore erosion, salinity interaction and loss of natural resources. Those measures could be structures such as dikes, levees, flood walls, sea walls, and dune building, and pollution control.

Economic Implications: While it is very difficult to estimate the economic implications correctly, it to possible to make some broad statements about the various options. In densely populated and productive areas retreat may not be viable in general and in the case of small islands, could be impossible. Accommodation could be extremely costly but could be justifiable in particular situation. Protection would be economically viable, in particular cases but could be unaffordable for poorer countries. One estimate quoted in the Inter-governmental Panel on Climate Change report gives the total additional costs to be more than 500 billion dollars with a major part to be incurred in developing countries.

Social and Cultural Implications: The lifestyles of people living on the coasts are distinct. The adaptive measures may, therefore impact heavily on the social and cultural mores of millions of people. Relocation of people whether planned or unplanned, could be very high and cause great conflicts, even within the same country. Where such relocation crosses national boundaries, international conflicts could occur. There is considerable optimism among many adoption to climate change is feasible, since it is as old as man himself. At every location, human beings and other living things adapted successfully to varying temperatures, precipitation, droughts, floods and storms. Because of economic and technological progress, the in-built adaptability of humankind has improved. The world has many climatic zones and as long as new climate in the area has existing analogues any where in the world people in that area can adapt and learn. Thus that is regarded as the best method through which a balance between the rising sea level and man can be maintained and the human species can see traces of survival for their future generation's comfortable and sound life.

Sea level rise is a problem, whose impact will be seen in nearly all the regions, especially coastal regions all over the world. But South Asian Region or the SAS is said to be the most vulnerable one. A vulnerability which is caused by factors like low lying coastal areas. Rapid deforestation, overgrowing rate of population, developing and poor economics and an uneducated population. These factors create all the more space for degradation of the coastal areas. Apart from the rise of sea level, other impacts of the global climate change, has left the South Asian Region in a very weakened position. The global warming which paves way for acid deposition, deforestation, forest firs, reduction of agricultures, and many other such kinds of hazards.

Global climate change and the rising sea level could also be responsible for the spread of diseases, an increase in their intensity, flooding of coastal areas, droughts, famine etc., a challenge that the South Asian Economics would not be in a position to meet. Concern about coastal swamps and marshlands comes from their special ecological value and the fact that they are already under stress from human development and pollution. And as these constitute an integral part of the South and South East Asian regions, especially in Bangladesh, Thailand, Indonesia, Sri Lanka, etc. where much of the land is bordered by the sea are in dire need of required steps to prevent the rising sea level, which if risen as predicted could severely hit their social and economic situation.

Special programmes thus have been started in the South Asian region. These programmes would not Just help combat the rise in the sea level but have also taken steps for prevention of the rise. Now it is widely accepted that the sea level is going to rise. Therefore, island nations and the third world under developed countries, which are situated in mountain belt and are likely to be most affected, should also examine the biological measures they can undertake to mitigate the adverse impact of sea level rise. Sea level rise (SLR) is a slow phenomenon, and its impact is gradual. It needs long term and precise observations to understand and predict the likely changes that may occur along the coasts of the continents and islands. These impacts of SLR are global in nature but they undergo annual, regional and local variations. They are influenced by the coastal morphology and rainfall, particularly in the countries situated in the monsoon belt on Asia.

In the delta and estuaries of rivers like Mekong in Thailand, Brahmaputra in Bangladesh, Ganges and Godavari in India, Indus in Pakistan, flooding is almost an annual feature and loss of human life and damage to property in coastal areas occur frequently. On the other hand, the coastline in the tropical countries is a preferred areas, as these regions are more productive and have a moderate climate and encourages fast development of large metropolitan cities as evidenced now in this zone. The coastline is being reclaimed and modified continuously. The coast of development and density of population is very high, therefore, if these areas get submerged due to climate change the loss will be very extensive.

Due to coastal process , hydrography and currents the continental margins, estuaries and islands undergo seasonal and long term fluctuations. The occurrence and disappearance of sandbars is caused due to silt movement. It influences closing and opening of the mouths of the estuaries and the coastal lakes and brings about changes in the hydrology and biological character of the area. Therefore, it is necessary to study the morphology of the coastal margins, estuaries, coastal lakes, back waters and lagoons and understand seasonal and annual variations. The discharge of sewage and organic wastes in estuaries undergo quantitative changes. In many parts, the estuaries get connected to low lying coastal plains, mud flats, and marshes and pollute them, thus large areas become unfit for development. In these areas, production of organic debris and detritus is a continuous process. The disposal and deposition of detritus and silt depends on coastal currents. Such areas are abundant near Bangkok, Calcutta, Bombay and in Shatal Arab in Iraq and are vulnerable to S.L.R.

Flooding changes the pattern of silt deposition, hydrography, rate of flow and mixing of estuarine, waters. Consequently, changes in biology, productivity, diversity of fauna and flora, reproduction and population composition are brought about. It is necessary to develop a comprehensive, long-term monitoring system using remote sensing, photo interpretation and verification by the ground trucks to study these areas. The ground truck requires continuous data updating, and data comparison, preparation of charts to catalogue annual changes. The study of long term changes will help in combating the increased sea level and its effect on the human beings in south Asia.

The developing countries of South Asia need a special treatment while talking of their prospects in the wake of the rising sea level and the climate change. Their economies need to be developed and only then can they dream of standing on par with the developed world of the North. As for a sustainable development this is an issue, which naturally must take priority with any nation and so the South Asian nations are no exception, but trying to force their growth to meet the conveniences of the North is quite an unreasonable requirement of the developed world. South Asia cannot be denied its right to grow nor to choose its own pathways to growth. Nor should that right be constrained by new conditions of trade or financial aid imposed in the name of environment, because who would know the meaning of the impact of the rise in sea level, better than south Asia? A fact that can't be denied is that the South Asian region does need financial support from the developed world, not just to develop, but to develop in the cleanest possible way.

The South Asian countries must be enabled to leap frog the environmentally destructive paths pursued in the west since the Industrial Revolution. To do this, they will need financial support, vastly improved professional and institutional, capacities and the cleanest and latest technologies.

Hiding behind the facade, of improving and developing in an environmentally clean way, the developed nations have somehow tried to curb South Asia's more gingerly taken steps in that direction, as was evident during the recently held Earth Summit at Rio de Janeiro, where, the Indian Minister for Environment Mr. Kamal Nath expressed his reservations against the treaty on climate change. In India's view, the 'safe' and 'clean' technology implied the exclusion of nuclear technology by which the US would like to bring in the controls on national polices through the backdoor as it were. This amounts to using environment to enforce the nuclear non-proliferation treaty. Thus this is not environment that is being talked of but something else altogether. This kind of an international political manipulation is what makes south Asia a part of the world that is disregarded of consideration of problem, which concerns it the most. What the North fails to understand is that if the South Asian Region had funds to develop economies in a sustainable fashion, why would it still risk its whole existence in development.

The problem of funding, which again was a major issue at the recently concluded Earth Summit in Rio is not just the North giving the South, monetary and technical help, but also securing its own future on this globe. There is a tendency to regard the costs of protecting the environment as an additional cost we can ill afford. But, much of what we often term 'cost' should really be regarded as investment which will produce a good return in economic as well as environmental terms. So the best way to move on to a sustainable growth would be through the help of the developed North. Environment is an issue for which even the individual of this earth is responsible for. So to allow policy makers to decide and rule the environment keeping in mind their own selfish interests is not part of a healthy scenario.

5 Ozone Depletion

The ozone layer is essential to life because it shields it from damaging ultraviolet radiation. Ironically, much less is known about the biological effects of increased ultraviolet radiation than about the chemical processes of ozone depletion in the atmosphere. Researchers are trying to learn how humans, vegetation, and aquatic ecosystems each may be affected by ozone depletion.

Scientists do know that direct exposure to ultraviolet radiation can damage the human immune system, cause cataracts, and increase the incidence of skin cancer. The EPA estimated in 1986 that the incidence of skin cancers would rise 2 per cent for each 1 per cent depletion of stratospheric ozone.

Today, mostly because of lifestyles that encourage skin exposure to strong sunlight, there are about 3000,000 to 400,000 new cases of skin cancer each year in the United States. As part of the effort to understand the effects on vegetation and crops, researchers have tested more than 200 plant species, two-thirds of which show sensitivity to increased ultraviolet exposure. Soybeans, one of civilization's staple food crops, is particularly susceptible to ozone damage, as are members of the bean and pea, squash and melon, and cabbage families.

Plant responses to ultraviolet radiation include reduced leaf size, stunted growth, poor seed quality, and increased susceptibility to weeds, disease, and pests. Scientists are also in the early stages of understanding how ultraviolet radiation might affect marine ecosystems and animals.

Concern about these systems begins with phytoplankton, microscopic marine algae that form the base of the marine food web. Studies in the tropics have shown that significant amounts of ultraviolet radiation can kill them, while lesser amounts can slow photosynthesis and thus productivity.

In Antarctica, this could affect kill, tiny crustaceans a notch up the food chain, and then fish, birds and marine mammals including seals and whales. While water provides some protection from radiation, crude estimates indicate that ultraviolet radiation can penetrate to depths of 10 to 20 meters. Some phytoplankton are known to be tolerant of ultraviolet radiation, whereas others cannot tolerate any. A likely response will be for tolerant species to replace sensitive ones, though no one knows how this would affect the fish that eat them.

The first unmistakable sign of human-induced change in the global environment arrived in 1985 when a team of British scientists published findings that stunned the world community of atmospheric chemists. Joseph Farman, of the British meteorological Survey, and colleagues reported in the scientific journal Nature that concentrations of stratospheric ozone above Antarctica had plunged more than 40 percent from 1960s baseline levels during October, the first month of spring in the Southern Hemisphere, between 1977 and 1984.

Most scientists greeted the news with disbelief. Existing theory simply had not predicted it. It meant that for several months of the year a hole forms in the ozone layer, which protects animals and plants from ultraviolet solar radiation. Suddenly it seemed that the chemical processes known to deplete ozone high in the earth's atmosphere were working faster and more efficiently than predicted. The discovery brought home a critical fact about the planet. No matter how much we learn about the workings of the earth system, the unexpected can always occur. Ground-based observations conducted by the National Oceanic and Atmospheric Administration (NOAA) since 1964 had not revealed the drop.

Measurements taken by the Total Ozone Mapping Spectrometer aboard the Nimbus 7 satellite operated by the National Aeronautics and Space Administration (NASA) since 1978 reflected the change but had not yet been analysed. When researchers scrutinized the data, Farman's findings were confirmed, but tough questions remained. What processes were causing the hole? Would the thinning of the ozone layer spread to other latitudes, or was it confined to the latitudes, or was it confined to the Antarctic.

To gather more information about Antarctic ozone chemistry and the ozone hole and its causes, a team of scientists led by atmospheric chemist Susan Solomon, of NOAA in Boulder, Colorado, headed off in 1986 on the first National Ozone Expedition to the Antarctic. By 1987 they and other teams of researchers had learned that the ozone over Antarctica had been reduced by more than 50 percent of values recorded in 1979, the first October of satellite operation, and that at altitudes between 15 and 20 kilometers, depletion was as great as 95 percent.

In 1988, temperatures (which influence processes in the stratosphere) were milder than in 1987, and in October the ozone declined by about 15 percent of 1979 values (already 20 percent below the baseline values of the 1960s). In 1989 temperatures dropped again, and ozone levels matched the severe depletion of 1987.

Until the hole was discovered, scientists were fairly sure that they understood the chemical processes at work in the ozone layer. Oxygen molecules (O_2), abundant throughout the atmosphere, are split apart into individual atoms (O + O) when energized by radiation from the sun. These atoms are free to collide with other O_2 molecules to form ozone (O_3).

The particular configuration of the ozone molecules allows them to absorb the sun's radiation in ultraviolet wavelengths that are harmful to life if they penetrate to the earth's surface. The ozone molecules formed by collision are partially removed by other naturally occurring chemical reactions, and so the overall concentration of stratospheric ozone remains constant.

High above the stratosphere, the density of gases is so low that oxygen atoms rarely find other molecules to collide with, and ozone does not form in abundance. Below the ozone layer, too title solar radiation penetrates to allow appreciable amounts of ozone to form. Thus most of the world's ozone is in a stratospheric layer bulging with ozone at latitudes from 10 to 35 kilometers.

Closer to the ground, in the troposphere, ozone produced through a series of chemical reactions involving hydrocarbons and nitrogen oxide emissions from vehicles and industrial activity is an effective greenhouse gas (in, addition to having adverse impacts on human health at high concentrations).

Thus ozone plays two very different roles in global environmental change: one in the stratosphere as a shield against harmful ultraviolet radiation, and another nearer the ground in the troposphere as a greenhouse gas and a health hazard. It is now known that in addition to the naturally occurring chemical reactions in the stratosphere, certain reactions involving chemical species of industrial origin, including chlorine and bromine, also chemically destroy ozone molecules.

Atmospheric chemists F. Sherwood Rowland, of the University of California at Irvine, and Mario J. Molina, now at Massachusetts Institute of Technology, first hypothesized the link between natural ozone in the stratosphere and chlorine released into the atmosphere from industrial sources. In 1973 they began to wonder: What happens to the industrially produced chlorinated molecules that are released into the lower atmosphere and for which no natural mechanisms for removal are known? The only long-lived natural source of chlorine in the earth's atmosphere is methyl chloride, which comes from the ocean and is present in the atmosphere at low levels.

The researchers hypothesized in 1974 that increasing concentrations of chlorofluorocarbons (CFCs), synthetic compounds that are chemically very stable in the lower atmosphere, rise unchanged through the lowest atmospheric layer, the troposphere. Even though CFCs are produced mostly in the industrialized countries of Europe and North America-where they are used in a wide variety of applications such as for solvents and refrigerants—they mix throughout the lower atmosphere, so that there are as many CFC molecules over Antarctica as over Colorado or Washington, D.C.

The researchers surmised that upon reaching the stratosphere, the CFCs encounter high-energy ultraviolet light, which breaks them down, releasing their chlorine atoms. The chlorine atoms can then engage with ozone in a catalytic reaction in which each chlorine fragment can destroy up to 100,000 ozone molecules before other chemical processes remove the chlorine from the atmosphere.

The hypothesis was borne out and improved by measurements and observations. In 1970 chlorine was present in the stratosphere at 1 2 parts per billion, and at about 3 parts per billion in 1985.

Were CFC use to continue at the 1985 rates (an eventuality precluded in 1987 by an international agreement known as the Montreal Protocol, described below), the stratosphere would contain about 8.2 parts per billion of total chlorine in the year 2050; current models of the chemistry and physics of the stratosphere suggest that at this concentration, total global ozone would drop by 5 per cent.

Rowland and Molina believed that most of the chlorine molecules that reached the stratosphere would form relatively inactive and harmless compounds. The ozone depletion would occur gradually, they hypothesized, and might not be detected for many years.

As ozone was lost, more ultraviolet radiation would reach the earth's surface. The researchers said two of the CFCs—CFC-11, which is widely used as a blowing agent in plastic foam, and CFC12, mostly used as a refrigerant—were particularly likely to destroy ozone because of their widespread use.

The two CFCs alone are increasing in the atmosphere at an annual rate of about 5 per cent. They are part of a class of chemicals known as halocarbons, many of which attack and destroy stratospheric ozone and also contribute to global warming as greenhouse gases.

Another chlorofluorocarbon, CFC- 113, is used as a solvent for cleaning electronic circuitry - Its atmospheric concentration is going up at an annual rate of about 11 per cent. Scientists are beginning to eye concentrations of still other synthetic halocarbons with suspicion. These include carbon tetrachloride, which is used as a cleaning fluid and in CFC production; methyl chloroform, used in solvents and adhesives; and halon 1301 and halon 121, which are used in fire extinguishers. Bromine, a chemical element that is related to chlorine and which is released from compounds used in fumigants and some fire extinguishers, is accumulating rapidly in the atmosphere. Bromine is believed to cause 10 to 30 per cent of the antarctic ozone depletion.

In the year before the discovery of the ozone hole, scientists were estimating that increasing use of chlorofluorocarbons might cause reductions in the total ozone at high latitudes by about one per cent in the 1980s and by 5 to 10 per cent 50 to 100 years from now. "While those numbers were disturbing," Solomon said, "they were nevertheless small enough that it was hard to argue that they were even real. They were not being observed." That viewpoint changed-along with scientists' faith in their models-in the mid-1980s as observations poured in from the coldest place on earth.

Now, many scientists describe the Antarctic ozone hole as the first unequivocal evidence of ozone loss due to man-made chlorine and one of the first clearly definable effects of human-induced global change When the Antarctic ozone hole was first discovered, little was known about the antarctic stratosphere beyond the ozone measurements themselves.

Virtually no data were available on the other chemical compounds present in the stratosphere, nor was there detailed meteorological information. This information was gathered rapidly by means of aircraft and state-of-the-art instrumentation.

In short order, scientists were able to measure a broad range of atmospheric compounds, including chlorine monoxide, chlorine dioxide, hydrochloric and nitric acid, nitric oxide and nitrogen dioxide, and nitrous oxide. They found that, the ozone levels dip at about the same latitudes where levels of chlorine monoxide ascend. As one researcher quipped, "These measurements are better than a smoking gun. This is more like seeing the guy pull the trigger."

Scientists can now calculate how much ozone would be lost with a given amount of chlorine monoxide. The answer is strikingly similar to the levels of zone depletion observed. Scientists are convinced that the elevated levels of chlorine and bromine account for much, if not all, of the Antarctic ozone depletion. For most of the year, the atmosphere over Antarctica has fairly high ozone concentrations.

The ozone molecules are formed over the tropics and are delivered along with chlorine to the Antarctic, as well as to the Arctic, via atmospheric motions. In Antarctica, a circulation pattern known as the antarctic polar vortex traps the ozone over the South Pole for several months. It is within this vortex that scientists have measured such shockingly low ozone concentrations during the first two weeks of October, shortly after the beginning of the Southern Hemisphere spring.

The explanation for the decrease /lies in the combination of ozone-destroying chemistry and weather conditions that favour formation of the highs, thin clouds known as polar stratospheric clouds (PSCs). The stratosphere is extremely dry, and the ice crystals that make up the clouds form only when temperatures drop to 80°C (- 112°F) or lower.

The clouds foster a basic change in stratospheric chemistry by allowing reactions to occur on surfaces rather than between gas molecules. The chemical reactions that take place on these surfaces convert chlorine from forms that do not react with ozone to other, less stable forms that readily break up in the presence of sunlight and go on to destroy ozone. Both cold temperatures and sunlight are critical to the process leading to ozone depletion in the Antarctic.

Antarctic ozone is depleted not during the winter, when temperatures are coldest and the South Pole is immersed in darkness, but in the southern spring, after sunlight returns but temperatures are still low. Researchers describe a process something like this: Usually chlorine in the stratosphere becomes trapped in so-called reservoir compounds, such as hydrogen chloride and chlorine nitrate, which themselves do not destroy ozone.

Once the stratosphere becomes cold enough that cloud particles freeze, the ice crystals provide surfaces on which reactions can occur: chlorine nitrate ($CLONO_2$) reacts with hydrochloric acid (HCl) present on the ice surface, producing molecular chlorine (Cl_2) and nitric acid (HNO_3). The nitric acid remains bound to the ice, and the molecular chlorine is quickly broken down into atomic chlorine (CI).

The chlorine atoms react with ozone (O_3), destroying it through the production of chlorine monoxide (ClO) and molecular oxygen (O_2) In a vicious cycle, the chlorine monoxide undergoes further reactions that reform a chlorine atom, which is then free to destroy another ozone molecule. As researchers improve their understanding of the antarctic ozone hole, it seems less ominous than it did at first for most of the rest of the world.

Over the mid-latitudes in the Southern Hemisphere, however, the hole may be spreading. Recent research suggests that in the late spring, when the antarctic vortex breaks up, the winds transport the polar, ozone-depleted air into lower latitudes.

The record low ozone values found over the Antarctic in October 1987 were followed by record low levels over Australia and New Zealand that December as the Southern Hemisphere summer began. NASA's Ozone Trends Panel reports that the effect may persist year round and that since 1979 ozone levels at all latitudes south of 60°S have decreased by 5 per cent or more.

For the most part, the hole has not spread outside of Antarctic and the lower Southern Hemisphere because it is limited by the seasons and the frigid temperatures required for the formation of the ice-laden polar stratospheric clouds. Yet, the insights gained during several years of intense data-gathering have raised concern about ozone in the stratosphere over the rest of the globe.

With ozone levels over the South Pole dropping up to 50 per cent or more for several months each year, scientists are eager to know whether the same processes are operating to deplete ozone over the Arctic. Results gathered by scores of atmospheric scientists using sensors aboard airplanes and balloons suggest that the arctic stratosphere differs from the antarctic stratosphere in a number of important ways that make a northern ozone hole of the same magnitude unlikely.

Measurements from satellites and ground-based stations reveal ozone losses of about 5 to 10 per cent at northern high latitudes during the arctic winter. This is much smaller than in the Antarctic for several reasons. For one, the arctic stratosphere generally warms up much earlier in the spring than does the antarctic, and the average temperatures are warmer. This means that cold temperatures and the sunlight necessary for the formation of polar stratospheric clouds-and the ozone depletion they promote-overlap for a much shorter interval. Another factor is that the arctic vortex is not as tight as the antarctic vortex.

As Rowland explains, air drifts across the pole, through the polar darkness, undergoes some polar stratospheric cloud chemistry, emerges into sunlight till in arctic winter-and loses a little ozone. Then the vortex warms up, and the ozone loss in the air mass stops. Meanwhile, another air mass is coming, and the process of successive small losses is repeated throughout the winter.

So far, the timing of the warming in the Arctic has offered some protection against wholesale ozone depletion. But researchers worry that this may not always be the case. In the winter of 1988-1989, the Arctic winter was unusually cold-the coldest for at least 25 years.

In January 1989 the polar vortex was relatively stable, giving rise to conditions similar to those in the antarctic winter stratosphere. In late January near the Artic Circle at Kiruna, Sweden, researchers measured an ozone deficit very similar to the initial stages of ozone depletion in early September in the Antarctic. With so much extra chlorine in the stratosphere, repeated occurrences of such winters could cause sudden ozone reductions over the Arctic and perhaps over much of the Northern Hemisphere.

Although the unusual chemistry of polar stratospheric clouds has made the Antarctic ozone layer more vulnerable than the rest of the atmosphere, there is the particularly troubling possibility that similar chemical reactions could occur in warmer latitudes. Temperatures outside the polar regions are 20° to 30°C too warm for ice clouds to form, but droplets of sulphuric acid and water can support reactions involving the same chlorine reservoir compounds that deplete stratospheric ozone over the Antarctic and may help to explain part of the 3 per cent ozone decrease observed over the Northern Hemisphere in the past two decades.

One prospect is that sulphurous particles emitted by a large volcanic eruption could team up with chlorine compounds to accelerate ozone destruction. Solomon and David J. Hofmann, of the University of Wyoming, describe a sharp drop in stratospheric ozone at mid-latitudes in 1982 after El Chichon erupted in Mexico, vaulting tons of volcanic debris into the upper atmosphere.

At the time, the ozone drop was unexplained; atmospheric chemists still thought in terms of gases, not surfaces of particles. In light of the recent ozone studies, it seems likely that the sudden increase in the availability of surfaces provided by the volcanic debris allowed the industrially produced chlorine compounds to break down into chlorine atoms that could then destroy ozone, though more slowly than in the Antarctic.

Many factors other than industrial chemicals affect the concentration of stratospheric ozone. Ozone ebbs and flows along with the cycle of sunspots. This solar cycle affects ozone because during the height of sunspot activity ultraviolet radiation increases at wavelengths that can split apart an oxygen molecule to form a molecule of ozone, causing a change of a few (91 to 2) per cent in ozone concentrations.

The solar cycle was winding down between 1979 and 1986, but it is currently increasing. The upswing in sunspot activity will lead to ozone production that could partially cancel the chlorine-caused decline, but this will be temporary. The researchers warn against a sense of false security: After 1991 ozone could decrease again. They also suspect that ozone responds to a 26 or 27-month cycle of varying wind direction in which shifts in winds from the equatorial stratosphere change the flow of ozone to the poles.

Still other factors fuel concern for the global ozone layer. Rowland and colleagues report that the amount of water in the normally and stratosphere could increase by 25 per cent by the middle of the next century (because of water vapour produced with oxidation of increasing amounts of methane in the atmosphere) and contribute to increased cloud formation.

The strong scientific consensus that CFCs deplete the ozone layer prompted nations to come together in unprecedented cooperation. The Montreal Protocol on Substances That Deplete the Ozone Layer, negotiated in September 1987, calls for a 50 per cent reduction in CFC production from 1986 levels by 1999. Forty-nine nations—including Canada, the United States Japan, and many nations in Europe, which together consume 80 per cent of the chemicals controlled—have ratified the protocol.

An important factor in the discussion leading to the protocol was the recognition that, because the chlorine compounds are so stable, CFC molecules emitted today will exist to deplete ozone for a century or more. The average lifetime of CFC-11, for instance, is believed to be about 75 years and for CFC-12, 110 to 140 years. With a 100-year average lifetime, Rowland explains, 37 per cent of the CFCs will still be in the stratosphere after 100 years, about 13 percent after 200 years, and about 4 percent after 300 years.

Researchers agree that CFC concentrations will continue to increase for 10 to 20 years after we stop releasing them to the atmosphere because they will escape from existing reservoirs such as automobile air conditioners, and because of the lag between emission, arrival in the high stratosphere, and decomposition.

Thus, if the nations that ratified the protocol comply with the terms established, average global ozone losses will still continue, but at a slower rate. These facts, and the growing body of scientific data on the threat to the ozone layer, are prompting nations to consider a 100 per cent reduction in CFC production by the year 2000.

The protocol is a delicate balance between the most up-to-date scientific information, reliable industrial expertise, and committed political leadership, all supported by strong and informed public interest. The Montreal Protocol may prove to be a model for actions that span national boundaries and interests as the world addresses common environmental issues such as green-house warming and other forms of global change. It is perhaps the best illustration of the emerging role of scientific information and scientists in discussions about policies to manage global change.

As Norway's former Prime Minister and chairperson of the World Commission on Environment and Development Gro Harlem Brundtland explains, "The scientist's chair is now firmly drawn up to the negotiating table, right next to that of the politician, the corporate manager, the lawyer, the economist, and the civic leader."Perhaps the most poignant image of our time is that of earth as seen by the space voyagers: a blue sphere, shimmering with life and light, alone and unique in the cold vastness of the cosmos. From this perspective, the maps of geopolitics vanish, and the underlying interconnectedness of all the components of this extraordinary living system-animal, plant, water, land, and atmosphere-becomes strikingly evident.

Humanity has learned that the activities of modern industrial economies, driven by consumer demands and burgeoning populations, can alter delicate natural balances. We can no longer pretend that nothing will happen as the planet is subjected to billions of tons of pollutants. It is not that Earth itself is necessarily fragile.

It may be rather, that our own tenure turns out to be less secure and less inevitable as a consequence of the planet's responses to unnatural conditions created by human actions. The Antarctic ozone hole conveyed a warning. Nature is capable of producing unpleasant surprises: even seemingly small interferences—in this case, an increase in stratospheric chlorine concentrations of a little more than one part per billion—can trigger dramatic and sudden reactions.

Recent experience with the forests of Europe and North America indicates that other areas may also have unforeseen thresholds beyond which natural processes are unable to absorb the assaults of contemporary society. The world may not have the luxury of early warning signals before an irreversible collapse occurs in some other part of the planet's ecosystem. A 1986 report by the National Aeronautics and Space Administration observed that "we are conducting one giant experiment on a global scale by increasing the concentrations of trace gases in the atmosphere without knowing the environmental consequences." This statement echoed almost verbatim the warning three decades earlier of Roger Revelle and Hans Suess, two scientists at the Scripps Institution of Oceanography.

Concerned about the implications for global warming of the rapid accumulation of carbon dioxide in the atmosphere resulting from fossil fuel combustion, they concluded in 1957 that "human beings are now carrying out a large scale geophysical experiment of a kind that could not have happened in the past. Worries over climate change have moved in recent years from farsighted members of the scientific community to the mainstream of international politics.

A consensus has developed that increasing atmospheric concentrations of carbon dioxide, along with other gases such as chlorofluorocarbons, methane, and nitrous oxide, could within a few decades cause temperatures to rise more rapidly, and to higher levels, than occurred in previous natural cycles going back for hundreds of thousands of years.

Such a development could have portentous implications, including widespread coastal inundation as a result of rising sea levels, major declines in agricultural productivity linked to changed rainfall patterns and increased soil erosion, greater incidence and severity of hurricanes and other storms, extinction of many plant and animal species, and creation of tens of millions of environmental refugees. There could be profound economic and social dislocations throughout the world.

Excessive greenhouse warming is an even more complex issue than depletion of the ozone layer. It involves more contributing factors, more wide-ranging and uncertain consequences, and more economically painful choices.

Modern society is faced with the prospect of rethinking many ways of doing business to which it has become accustomed. Efforts to limit the magnitude and rate of temperature rise, and to adapt to its effects, will undoubtedly imply changes in energy, industry, agriculture, development, and population policies, as well as in consumption habits. The new environmental threats to national and planetary security—of which climate change appears to be the most far-reaching—

challenge both traditional science and diplomacy. A new science has evolved in recent years, made possible by advances in computer modelling, satellites, and measurement technologies.

Known as earth systems science, this discipline attempts to integrate chemistry, physics, biology, geology, anthropology, meteorology, oceanography, and other subjects in order to understand more fully the interrelated forces that govern this planet. In 1986 the International Council of Scientific Unions (ICSU) launched a long-term interdisciplinary initiative that has been described as the biggest international scientific effort ever organized.

Its purpose is to develop new insights into plentary processes and the ways they are being affected by human activities. Designated as the International Geosphere-Biosphere Programme, it was adopted unanimously by the more than 70 national academies and 20 international scientific unions that form the membership of ICSU—a manifestation of universal concern about the seriousness of the problems.

Because cooperation among sovereign states is essential for developing effective policies to address these issues, the new science requires, an analogue in the realm of international relations., The negotiators of the Vienna Convention and the Montreal Protocol faced issues similar to those raised by potential climate change. The science was uncertain, and the predicted harmful effects, though grave, were remote and unproved.

Entrenched industrial interests claimed that new government regulations would cause immense economic and social dislocations. Technological solutions were either no inexistent or were considered unacceptable by most major participating governments.

Under these conditions, some governments allowed commercial self-interest to influence their scientific positions and used the scientific uncertainty as an excuse for delaying difficult decisions.

Many political leaders were long prepared to accept future environmental risks rather than to impose the certain short-term costs entailed in limiting use of products seen as necessary to modern standards of living. Short-range political and economic concerns were thus formidable obstacles to cooperative international action based on the ozone-depletion theory.

Government policymakers face a dilemma in attempting to deal with these new environmental challenges. Premature actions or regulations based on imprecise and possibly incorrect theories and data can incur costs that later turn out to be unnecessary. But postponing a decision also might not be cost free. Waiting for more complete evidence can run the risk of acting too late to prevent major and possibly irreversible damage. In this eventuality, the future economic and social costs could be much higher, perhaps even catastrophic.

Unfortunately, the current state of economics is not helpful in analysing such a situation. Its traditional methods of measuring income and growth appear increasingly irrelevant in the modern world: the more chloroflurocarbons a country produces, for example, the greater is the growth in its gross national product. And the economist's prescription for coping with future monetary flows is inadequate for responding to large but distant dangers: present-value discounting can reduce even huge long-term costs to insignificance. Thus, the application by policymakers and investors of the tools of conventional economics may result in ecologically very misleading decisions.

The history of the efforts to protect the ozone layer clearly demonstrates the crucial role played by industry in developing and implementing international environmental policy.

Although US initiatives were instrumental in first reducing CFC emissions in the 1970s, in achieving the Montreal Protocol in 1987, and in strengthening it in 1990, on four occasions during the ozone negotiations the U.S. government either reversed or came close to reversing its position. It is noteworthy that on the three occasions since 1985, pragmatically oriented American industry forces intervened in favour of the international regulatory regime.

The response of industry to an environmental problem is conditioned by a complex of considerations and pressures generated by the market, which is itself directly influenced by consumer preferences and by government regulatory actions; and underlying these immediate influences may be educational activities of the media and environmental organizations.

The personal value of corporate leadership and stockholders, as well as general societal attitudes toward a given environmental issue, will also affect industry reactions. In the early years of the ozone history, both American and European industrialists were resolutely opposed to controls over CFCs. It was almost as if business leaders simply could not bring themselves to believe that these seemingly ideal chemicals, with so many benefits to society, could be capable of inflicting a remote outrage on the environment.

American industry, however, through the Chemical Manufacturers Association, was consistently committed to resolving the uncertainties raised by the scientists, even if the results were to prove unpleasant; industry strongly promoted international scientific research and the 1985 Vienna Convention. The position of American industry toward international regulation was also conditioned by the desire for a: "level playing field" with their European competitors.

During the negotiations before both the Montreal conference in 1987 and the London meeting of parties in 1990, industrialists almost inevitably complained that the proposed controls were too harsh and too costly to implement, while environmental groups argued that they did not go far enough.

The treaty negotiators had to seek a balance between being too soft on industry and thereby running greater risks to the ozone layer, and demanding too much. Economically or technically unrealistic controls could generate severe dislocations. Bitter industry resistance, loss of jobs, long court battles to delay implementation, consumer backlash, non compliance, could all delay the needed structural changes and substitute products.

Industry needs a firm signal and a stable regulatory environment to give it the confidence to invest in new and safer technologies; this consideration emerged very clearly in the debate during 1989 and 1990 over the HCFC substitute chemicals. Sometimes, however, it may be necessary to overcome severe inertia or preoccupation with near-term profits.

Control measures may, therefore, need to be more stringent than industry would like, in order to provide a spur to environmentally responsible investment. As West German environment minister Klaus Topfer (a member of the conservative Christian Democratic party) put it in June 1990: "1 am absolutely convinced that if you give a clear cut timetable, it will stimulate industry to come up with substitutes. But if more time is allowed. I really believe they will take more time." In these circumstances, informed consumers can have a decisive influence. In the case of CFCs, aroused American consumers, through their changed purchasing habits, brought down the U.S. aerosol spray market, by two-thirds in the late 1970s, and thereby stimulated competitive development of substitutes even before government regulations were in place.In 1988, threatened consumer boycotts against CFC aerosols in the United Kingdom, promoted by environmental groups, convinced British producers to phasc out these products. Citizens of Taiwan blocked the erection of a CFC production facility in 1989. Public opinion in the Federal Republic of Germany had reached such a state by 1990 that it created the conditions for the earliest phaseout of ozone-depleting substances by any major producing country-with the voluntary cooperation of private industry.

Even though large companies have substantial resources available for public relations and advertising, it is no longer worth while for them to try to resist when consumer feelings become sufficiently intense. And this same public interest can help to discourage possible temptations to evade environmental regulations, whether by exploiting loopholes or by moving to other countries.

Unfortunately, amid the cacophony of competing claims and data, there are no ready formulas for government policymakers to apply in achieving the right balance of risk insurance. In the ozone negotiations, tentative proposals were advanced, analysed, and debated with both industrialists and environmentalist. In the end, the negotiators had to rely on some combination of experience and instinct when it came to drawing the line. That neither industry nor the environmental groups were ever entirely satisfied with the results is perhaps inevitable and may even be an indicator of a reasonable outcome.

The international community was successful in its approach to the problem of protecting the stratospheric ozone layer. The experience gained from this process suggests several elements of a new kind of diplomacy for addressing such similar global ecological threats as greenhouse warming.

1. *Scientists must play an unaccustomed but critical role in international environmental negotiations.* Without modern science and technology, the world would have remained unaware of the danger to ozone until it was too late. Science became the driving force behind ozone policy. The formation of a commonly accepted body of data and analyses and the narrowing of ranges of uncertainty were prerequisites to a political solution among negotiating parties initially far apart.

   In effect, a community of scientists from many nations, committed to scientific objectivity, developed through their research an interest in protecting the planet's ozone layer that transcended divergent national interests. In this process, the scientists had to assume some responsibility for relating the implications of the findings to alternative remedial strategies. Close collaboration between scientists and key government officials who also became convinced of the long term dangers ultimately prevailed over the more parochial and short-run interests of some national politicians and industrialists.

2. *Governments may have to act while there is still scientific uncertainty, responsibly balancing the risks and costs of acting or not acting.* By the time the evidence on such issues as the ozone layer and climate change is beyond all dispute, the damage maybe irreversible, and it may be too late to forestall serious harm to human life and draconian costs to society.

Politicians must therefore resist a tendency to lend too much criedence to self-serving economic interests that demand scientific certainty, maintain that dangers are remote and unlikely, and insist that the costs of changing their ways are astronomical. The signatories at Montreal knowingly imposed substantial short-run economic dislocations even though the evidence was incomplete; the prudence off their decision was demonstrated when the scientific models turned out to have underestimated the effects of CFCs on ozone. Governments must sponsor the needed research and act responsibly on the basis of often equivocal results. Unfortunately, the current tools of economic analysis are inadequate aids in this task and can even be deceptive indicators; they are in urgent need of reform.

3. *Educating and mobilizing public opinion are essential to generate pressure on hesitant governments and private companies.* The interest of the media in the ozone issue and the collaboration with television and press by diplomats, environmental groups, and legislators had a major influence on governmental decisions and on the international negotiations. Concerned consumers brought about the collapse of the CFC aerosol market. And in their educational efforts, the proponents of ozone layer protection generally avoided invoking apocalypse and resisted temptations to overstate their case in order to capture public attention; exaggerated pronouncements could have damaged credibility and provided ammunition to those interest groups that wanted to delay action.

4. *Multilateral diplomacy, involving coordinated negotiations among many governments, is essential when the issues have planetary consequences.* The manifold activities of an international organization-the United

Nations Environment Programme-were crucial in promoting a global approach to the protection of stratospheric ozone. UNEP coordinated research, informed governments and world public opinion, and played an indispensable catalytic and mediating role during the negotiation and implementation of the protocol.

5. *Strong leadership by a major country can be a significant force for developing international consensus.* The U.S. government reflected its concerns over the fate of the ozone layer through stimulating and supporting both American and international scientific research. Then, convinced of the dangers, it undertook extensive diplomatic and scientific initiatives to promote an ozone protection plan to other countries, many of which were initially hostile or indifferent to the idea. As the largest emitter of both ozone-destroying chemicals and greenhouse gases, the United States has enormous potential to influence the policy considerations of other governments in favour of environmental protection. In fact, because of the geographic size and population of the United States, its economic and scientific strength, and its international interests and influence, progress in addressing global environmental problems can probably not be achieved without American leadership.

6. *It may be desirable for a leading country or group of countries to take pre-emptive environmental protection measures in advance of a global agreement.* When influential governments make such a commitment, they legitimate change and thereby undercut the arguments of those who insist that change is impossible. Pre-emptive actions can also support moral suasion in encouraging future participation by other countries. In addition, action by major

countries can slow dangerous trends and hence buy time for future negotiations and for development of technological solutions. The 1987 US ban on aerosols relieved pressure on the ozone layer and lent greater authority to the US government when in subsequently compaigned for even more stringent worldwide measures. Although environmental controls might conceivably affect a country's international competitiveness in the short run, they might also, by stimulating research into alternative technologies, give that country's industry a head start on the future.

7. *The private sector-including citizens groups and industry-is very much involved in the new diplomacy.* The activities of both environmental organizations and private industry in undertaking research, lobbying governments, and influencing public opinion definitely affected the international debate on the ozone issue. A major by product of the ozone negotiations was the development of closer relations among hitherto separate environmental groups around the world, reflected in their new and unaccustomed co-operation at conferences on the ozone layer and climate change in 1989 and 1990. Environmental organizations can also play an informal future watchdog role in monitoring compliance by industry with internationally agreed-upon commitments. For their part, industrialists are becoming aware that their corporate image is increasingly affected by environmental issues. The intellectual and financial resources of the corporate sector are, moreover, essential for developing the required technological solutions. These groups will expect to be involved in establishing policies and negotiating international accords on climate and other emerging global ecological issues.

8. *Economic and structural inequalities among countries must be adequately reflected in any international regulatory regime.* In the longer run, the developing countries, with their huge and growing populations, could undermine efforts to protect the global environment. As a consequence of the ozone issue, the richer nations for the first time acknowledged a responsibility to help developing countries to implement needed environmental policies without sacrificing aspirations for improved standards of living. The Montreal Protocol broke new ground with the creation of the multilateral ozone fund and the commitments for technology transfer, while illuminating the issues that must be considered in arriving at realistic and equitable solutions to future global problems.

9. *The effectiveness of a regulatory agreement is enhanced when it employs market incentives to stimulate technological innovation.* Technology is dynamic and not, as some industrialists have seemed to imply, a static element. But left completely on its own, the market does not necessarily bring forth the right technologies to protect the environment. The ozone protocol set targets that were initially beyond the reach of the existing best-available technology. They appeared difficult but were in fact achievable for most of industry-and thereby averted monolithic industrial opposition that might have delayed international agreement. The treaty actually stimulated collaboration among otherwise competing companies in research and testing, thereby saving both time and money in the development of replacement technologies and products. By expeditiously getting the protocol established in international law even with a 50 per cent reduction target, the negotiators effectively

signalled the marketplace that research into solutions would now be profitable-thus setting the stage for the later decisions for phaseout.

10. *The signing of a treaty is not necessarily the decisive event in a negotiation; the process before and after signing is critical.* It was extremely important to separate the complicated ozone-protection issue into manageable components. The informal fact-finding efforts during 1985 and 1986 and again during 1989 and 1990-workshops, conferences, consultations—established an environment conducive to building personal relationships and generating creative ideas, and thereby facilitated the formal negotiations. During the negotiations themselves, the use of small working groups and a single basic "chairman's text" aided in the gradual emergence of consensus. The developments following the 1987 signing illustrated the wisdom of designing the treaty as a flexible instrument. By providing for periodic integrated assessments—the first of which was advanced from 1990 to 1989 in response to the rapidly changing science-the negotiators made the accord adaptable to evolving circumstances. In effect, the protocol became a continuing process rather than a static solution.

11. *Firmness and pragmatism combined are important ingredients of diplomatic success.* The proponents of strong controls refrained from extreme positions but never relented in their pressure for a meaningful treaty. They did not insist on perfect solutions that might have unnecessarily prolonged the negotiations. Nor did they wait for universal participation, or even for agreement among all potential major players. Instead, they achieved an interim solution with built-in flexibility that could serve as a springboard for future action.

12. *Individuals can make a surprisingly significant difference.* UNEPs Mostafa Tolba provided overall personal leadership, initiating critical consultations with key governments, private interest groups, and international organizations. During the negotiations, he moved from group to group, arguing for flexibility, applying pressure, often floating his own proposals as a stimulus to the participants. Individual scientists, negotiators, environmentalists, and industry officials also provided ideas, decisions, and actions that proved vital to the final outcome.

The relevance of the experience with the ozone treaties has not been lost on the international community as it addresses greenhouse warming. The establishment in late 1988 and the subsequent functioning of the Inter-governmental Panel on Climate Change—with its multiple scientific, economic, and policy workshops and varied participation from public and private sectors—parallel the fact-gathering and analytical phases of the ozone history.

UNEP and the World Meteorological Organization are again actively involved in the process. Similarly, many governments announced their support during 1989 for a framework agreement on climate change, comparable to the 1985 Vienna Convention on Protecting the Ozone Layer. Such a climate convention need not be a complicated undertaking, and it should be achieved at the earliest possible date. The existence of gaps in scientific and economic knowledge should not become an excuse for delaying the negotiations.

Ideally, a framework convention would enable governments to formalize their agreement in principle on the dimensions of the climate problem and the scope of possible responses. Governments would undertake general obligations for policies and actions to mitigate and adapt to global warming. They would also agree on coordinated research and monitoring to develop additional data as guidelines for future measures. It might be useful to go beyond the Vienna precedent and try to build into a climate convention some general targets and timetables.

However it would probably be questionable for advocates of stringent greenhouse gas controls to attempt to load a convention with overly detailed and still controversial commitments. Premature insistence on optimal solutions could have the unintended effect of bogging down the negotiations and prolonging the entire process. On the other hand, an early convention—even if general in its terms—would itself establish apolitical momentum for further concrete actions.

The framework convention would provide the legal and logistical structure for the crucial step corresponding to the Montreal Protocol: agreement on specific international measures. Work on such measures could well begin even before completion of the framework convention. Because of the complexity of the climate issue, it would not be realistic to attempt to achieve an ideal set of responses at a single stroke; here again, the quest for perfection might only serve to delay action.

Instead, it might be more effective-and realistic-to think in terms of incremental stages and partial solutions, following the examples of the Vienna Convention and the Montreal Protocol. Thus, governments could negotiate a number of separate implementing protocols, each one containing specific measures for dealing with a different aspect of the climate problem. One example would be a treaty mandating greater energy efficiency in the transportation sector, which should be feasible because it need involve only a relative handful of major producers of transportation equipment.

The ozone accord itself exemplifies interim progress on the climate problem by means of a constituent protocol. Such partial solutions can be significant; a NASA study estimated that if CFCs had continued to increase at the growth rates of the 1970s, they would by 1989 have surpassed carbon dioxide in their greenhouse impact. It might be useful to establish standing negotiations under a permanent secretariat, similar to arrangements for the Geneva disarmament talks. By this means, individual protocols could simultaneously be in process of development, each at its own pace. A climate convention and protocols need not strive right away for universal membership-that could be an unnecessarily complicating factor.

In actuality, the overwhelming proportion of carbon emissions from fossil fuels and deforestation is concentrated in a relatively small number of industrialized and developing nations. Indeed, the major industrialized countries, which are primarily responsible for the world's current precarious ecological condition, could make a vital contribution by agreeing on pre-emptive actions even before a broader climate treaty is negotiated.

The United States and Canada, the Soviet Union, the European Community, and Japan together account for about 60 per cent of carbon emissions from fossil fuels. By not delaying actions to increase energy efficiency and reduce carbon dioxide emissions, these countries could significantly slow the warming trend. Doing this would buy time for innovation in energy efficient technologies and renewable energy sources that could be shared with developing and Eastern European countries to aid them in assuming their own responsibility.

6 PHOTOCHEMICAL SMOG IN THE ATMOSPHERE

Photochemical oxidants are secondary pollutants producing in the atmosphere form reactions involving primary pollutants. Both groups will be described together because the majority of photochemical oxidants originate from reactions in which hydrocarbons are directly or indirectly involved. Any estimate of the effects of atmosphere hydrocarbon pollution must include the contributions made by phtochemical oxidants. In fact, most of the principal effects of hydrocarbon pollution are brought about by compounds producing from atmospheric reactions of hydrocarbons. The number of hydrocarbons involved in air pollution is fairly large.

An analysis of urban air by gas chromatography has resulted in the identification of 56 different hydrocarbon compounds. The number present is probably large, but the sensitivity of the analytical method prevented others, in very low concentrations, from being detected. Natural sources contribute huge quantities of these substance. Plants, particularly trees, emit hydrocarbons.

Most of the hydrocarbons emitted to the atmosphere by plants are thought to be *terpanes*. Most of the plants which produce terpenes belong to the family *Coniferae*, the family *Myrtaceace*, and the genus *Citrus*. Isoprene (2-methyl-1, 3-butadiene), a hemiterpene, has been identified in the emissions of cottonwood, eucalyptus, oak, sweetgum, and white spruce trees. Turpentine, common terpene, is widely used in paint because it reacts with atmospheric oxygen to form a peroxide, then a hard resin. It is likely that compounds such as a-pinene and isoprene undergo similar reactions in the atmosphere to form particulate matter.

The resulting Aitken nuclei aerosols probably cause the blue haze associated with the atmosphere above some heavy growths of vegetation. In addition to the hydrocarbons produced by higher plants, methane is the other major naturally occurring hydrocarbon emitted to the atmosphere. Methane is formed in huge quantities by bacteria in the anaerobic decomposition of organic matter in water, sediments, and soil.

$$2CH_2O \xrightarrow{\text{bacteria}} CO_2 + CH_4 \qquad \text{...(1)}$$

Most of the pollutant hydrocarbons added to the atmosphere through the activities of man are produced by automobiles. Global sources of hydrocarbon emissions are estimated in Table 1. The net air pollution effect of a given source of hydrocarbons is determined by the overall reactivity of hydrocarbons from the source as well as total amount. Most hydrocarbon pollutant sources produce about 15 per cent reactive hydrocarbons, whereas those from gasoline are about 45 per cent reactive.

A number of different specific hydrocarbon compounds are found in regions heavily polluted by such substances. Over 20 different compounds were identified and analyzed. Of these, methane was present at by far the highest concentration, an average of 2400 *parts per billion* (ppb) by volume in downtown Los Angeles.

Aside from methane, 80% of the remaining hydrocarbons consisted of 10 compounds. These compounds and their average levels were ethane and ethylene (102 ppb total), acetylene and propane (76 ppb total), *n*-butane (46 ppb), isopentane (35 ppb), toluene (30 ppb), *n*-pentane (21 ppb), *m*-xylene (12 ppb), and isobutane (12 ppb).

Table 1
Global and U.S. Emissions of Pollutant Hydrocarbons

Source	Emissions
Petroleum	%
Refining	7.2
Oils and distillates	0.4
Gasoline	38.5
Evaporation in transfer and storage	8.8
Coal	
Heating	2.3
Power generation	0.2
Industrial uses	0.8
Wood (fuel and forest fires)	2.2
Incinerators and refuse burning	28.3
Solvent evaporation	11.3

It was observed that as the day progressed the concentration of unsaturated hydrocarbons (butenes) decreased to a greater extent than that of the paraffins. The decrease in the paraffins was attributed entirely to mixing and consequent dilution. Industrial processes constitute the next largest source. Hydrocarbons get lost to the atmosphere during production, processing storage, and transfer of products containing them. Gasoline hydrocarbons are lost in vapour form during the loading of tank trucks, the filling of service station storage tanks, and the filling of service station storage tanks, and the filling of automobile fuel tanks. The evaporation of organic solvents, most of which contain hydrocarbons, accounts for nearly 10% of all industrial emissions. These solvents are important ingredients of paints, varnishes, lacquers, undercoatings and other similar products. These products consist of 40.80% solvent, which usually evaporates during or after application. The processes involved in the manufacture of these products also comprise potential hydrocarbon sources.

Pollution due to Automobilies

The gasoline powered automobile is the major culprit in producing pollutant hydrocarbons. At the high temperatures and pressures encountered in the automobile engine, gasoline which is not completely burned undergoes chemical reactions which produce several hundred kinds of hydrocarbons.

Some of these are highly reactive in the atmosphere, and others, such as benzopyrene, are known carcinogens. These hydrocarbons spew from automobile exhausts into the faces of pedestrians and drift upward in the atmosphere where, with the help of solar energy, they are cooked into an even more exotic mixture of photochemical oxidation products above our cities. It is little wonder, therefore, that the control of automotive hydrocarbons has had the top priority in the effort to acquire clean air. Aside from the exhust, the automobile contains several potential sources of pollutant hydrocarbons. One source is the crankcase, which is having mist composed of lubricating oil, and receives "blowby" consisting of exhaust gas and unoxidized carbureted mixture from around the pistons. Blowby is recirculated through the engine intake manifold by way of the PCV (positive crankcase ventilation) valve. These emissions were the first to be controlled. The fuel system has been a second major source of automotive hydrocarbon emissions. Fuel may be given off through fuel tank and carburetor vents. When the engine is shut down and engine heat worms up the fuel system, gasoline may be evaporated and emitted to the atmosphere. In addition, hearing during the daytime and cooling at night causes the fuel tank to breathe and emit gasoline fumes. Many systems have been devised for the control of evaporative emissions from fuel systems. One is a carbon canister which is able to collect evaporated fuel from the fuel tank and fuel system, to be purged and burned when the engine is operating. Reduction of fuel volatility also helps cut down hydrocarbon evaporation from fuel systems.

The automobile exhaust is the greatest source of hydrocarbons. They are produced by the engine, which for the vast majority of automobiles is a fourcycle internal combustion engine fueled with gasoline. The four basic steps in the complete cycle of the engine are the following:

1. *Intake*. An air-gasoline mixture formed in the carbouretor and consisting of about 14-15 parts by weight of air per part gasoline gets drawn into the cylinder through the open intake valve.

2. *Compression*. The combustible mixture gets compressed by a ratio of about 8.5 :1. Higher compression ratios tend to favour thermal efficiency and complete combustion of hydrocarbons. However, higher temperature, premature combustion ("pingin"), and high production of nitrogen oxides also result from higher combustion ratios.

3. *Ignition and power stroke.* As the fuel-air mixture gets ignited by a spark plug near top-dead-center of the combustion stroke, a temperature of about 2500°C is reached very rapidly at pressures up to 40 atm. The temperature decreases in a few milliseconds. This rapid cooling "freezes" nitric oxide in the form of NO without permitting it time to dissociate to N_2 and O_2, thermodynamically favoured at the temperature and pressure of the atmosphere.
4. *Exhaust.* Exhaust gases having largely of N_2 and CO_2 with trace of CO, NO, hydrocarbons, and O_2 get pushed out through the open exhaust valve, thus completing the cycle.

The primary cause of hydrocarbons in the exhaust gas has been "wall quench." The relatively cool wall in the combustion chamber of the internal combustion engine quenches the flame within several thousandths of an inch of the wall. Part of the remaining hydrocarban may get retained as residual gas in the cylinder, and part may get oxidized in the exhaust system. The remainder goes to the atmosphere as pollutant hydrocarbons. Engine misfire due to the improper adjustment and decleration greatly tends to increase the emissions of hydrocarbons. Turbine engines are not subject to the wall quench phenomenon because their surfaces are always hot. Several engine design characteristics favour low exhaust hydrocarbon emissions.

Spark retard reduces exhaust hydrocarbon emissions. One reason for this reduction is that the effective surface/volume ratio of the combustion chamber is reduced. Secondly, when the spark is retarded, the combustion products are purged from the cylinders sooner after combustion. Therefore, the exhaust gas is hotter, and reactions consuming hydrocarbons get promoted in the exhaust system.

The air/fuel ratio in the internal combustion engine has a marked effect upon the emission of hydrocarbons. As the air/fuel ratio becomes richer in fuel than the stoichiometric ratio, the emission of hydrocarbons increases sharply. The "air pump" or "Thermactor" is used to control hydrocarbon as well as carbon monoxide emissions.

A more sophisticated device similar to the Thermactor is the thermal exhaust reactor which is insulated and provides a large volume in which exhaust gases can be retained for some time to react at high temperature with fresh air pumped into the system.

Many catalytic exhaust reactors have been devised for the oxidation of excess hydrocarbons. For the control of hydrocarbons and carbon monoxide, these catalytic reactors require an excess of oxygen, whereas for the removal of NO_x a reducing atmosphere is required. Therefore, two beds of catalyst are required.

Catalytic exhaust reactors using platinum catalysts came into widespread use in the United States during the 1975 model year. Since lead poisons the catalyst used, the automobiles equipped with catalytic exhaust control devices require lead-free gasoline, and facilities for dispensing this fuel have been installed in all major service stations.

Recent evidence also suggests that catalytic exhaust reactors result in the production of potentially hazardous levels of sulfuric acid by catalysis of the oxidation of sulfur dioxide from sulfur-containing gasoline:

$$2SO_2 + O_2 \rightarrow 2SO_3 \quad ...(1)$$

$$SO_3 + H_2O \rightarrow H_2SO_4 \quad ...(2)$$

Photochemical Oxidants

The term photochemical oxidant is used to describe an atmospheric substance, produced by a photochemical process (a chemical process that requires light), which will oxidize materials not readily oxidized by gaseous oxygen. These substances are secondary pollutants produced by an interaction of primary pollutants with light. Hydrocarbons are very much involved in the production of photochemical oxidants. The large variety of hydrocarbons involved in these reactions prevents a simple explanation from being formulated for the various processes.

Although the intimate details of the reactions are not known, several specific facts are known quite well. The involvement of the NO_2 photolytic cycle has been definitely established. Two of the most harmful secondary pollutants produced as a result of the involvement of hydrocarbons in this cycle have been identified. Ozone is obviously not a hydrocarbon, but its concentration increase in the atmosphere is a direct result of hydrocarbon reaction PAN is seen to be a hydrocarbon derivative.

Formation of Photochemical Oxidants

The majority of the harmful effects of hydrocarbon pollution are not caused by the hydrocarbons themselves, but by the products of photochemical reactions in which the dydrocarbons are involved. Hydrocarbons do not readily react with sunlight, but they are reactive toward other substances produced photochemically. The photolytic cycle of NO generate several material with which hydrocarbons readily react. The NO_2 cycle involves the three reactions:

$$NO_2 + \text{Sunlight} \rightarrow NO + O \quad \text{...(1)}$$

$$O + O_2 \rightarrow O_3 \quad \text{...(2)}$$

$$O_3 + NO \rightarrow NO_2 + O_2 \quad \text{...(3)}$$

and the net effect should be a rapid cycling of NO_2. Ozone (O_3) and NO should be formed and destroyed equal quantities. The cycle gets distrupted when hydrocarbons are present because of their ability to react with either the oxygen atoms (O) produced (equation 1) or the ozone produced (equation 2). The most likely reaction is that between O and hydrocarbons, because it takes place 10^8 times faster than the one between O_3 and hydrocarbons. The reaction between O and O_2 is faster than that between oxygen atoms and hydrocarbons, but the latter reaction still proceeds fast enough to upset the cycle. The product of a reaction between O and a hydrocarbon is a very reactive intermediate species called a hydrocarbon free radical (RO_2). Free radicals of this type can further react with a number of different species including NO, NO_2, O_2 O_2, O_3, and other hydrocarbons. Some general facts are known.

1. The free radicals react rapidly with NO to produce NO_2. The consequence of this is that with NO removed from the cycle, the normal mechanism for O_3 removal (equation 3) has been eliminated, and the concentration of O_3 in air gets increased.
2. The free radicals can react with O_2 and N_2 to give peroxyacylnitrate (PAN).
3. The free radicals can react with other hydrocarbons and oxygen species to produce additional undesirable organic compounds.

The mixture of products resulting from the hydrocarbon interference in the NO_2 photolytic cycle is called photochemical smog. It consists of occumulated O_3, CO, PAN, and other organic compounds, including aldehydes, ketones, and alkyl nitrates.There is a wide variation in the tendency for different hydrocarbons to enter into the photochemical air pollution process. As a consequence, it is impossible to predict accurately, on the basis of total hydrocarbon concentration, the contribution of a hydrocarbon mixture to photochemical air pollution. The detailed composition of the mixture must be known.

Importance of Hydrocarbon Pollution

Except in cases of very severe contamination, generally involving polycyclic aromatic hydrocarbons, atmospheric hydrocarbons in themselves are not severe air pollutants. However, hydrocarbons entering the atmosphere as primary pollutants produce photochemical oxidants as secondary pollutants. These oxidants give rise to a number of undesirable effects such as the notorious smog which afflicts Los Angeles and other urban areas. Atmospheric oxidants cause eye and respiratory tract irritation and are suspected of imparing human health.

Some plants have been very susceptible to damage by oxidants. One such oxidant is ozone, O_3. Chemically, ozone is of course entirely different from the hydrocarbons which are required for its formation in low level atmospheres. Other oxidants like peroxyacetyl-nitrate, PAN, differ a great deal chemically from their hydrocarbon precursors. However, because of the close relationship between pollution and photochemical oxidant formation, hydrocarbons and oxidants are discussed together.

Reactions of Hydrocarbons in the Atmosphere

Hydrocarbons undergo chemical and photochemical reactions in the atmosphere. With carbon monoxide as a nonreactive standard for pollutant emissions, the expected levels of hydrocarbons in the atmosphere may be calculated from the known composition of pollutant sources. The photochemical reactions are able to initiate most of the chemical processes culminating in removal of hydrocarbons from the atmosphere. The most important of these primary photochemical reactions in an atmosphere is

$$NO_2 + hv \rightarrow NO + O \qquad ...(1)$$

whereby nitrogen dioxide undergoes photochemical dissociation to yield NO and atomic oxygen. The second most important primary photochemical reaction probably involves the formation of free radicals from aldehydes:

$$CH_3 - \overset{\overset{\displaystyle O}{\|}}{C}\text{-}H + hv \rightarrow CH_2 + H - CO \qquad ...(2)$$

It is possible that in some atmospheres other primary photochemical processes may be more important than the above. In addition to aldehydes, compounds forming free radicals in the atmosphere include ketones, organic peroxides, organic nitrates, and hydrogen peroxide. Free radicals certainly are also involved in the reactions of atmospheric hydrocarbons. For example, it has been hypothesized that the initial reaction of *n*-butane by photolysis in an atmosphere containing NO_2 is the reaction the paraffin with atomic oxygen [produced by Reaction (1)].

$$C_4H_{10} + O \rightarrow C_4H_9 + OH \qquad ...(3)$$

Free radicals may react with other species to form other radicals. For example, the *n*-butyl radical formed in Reaction (3) is thought to react with molecular oxygen to from a peroxy radical:

$$C_4H_9 + O_2 \rightarrow C_4H_9OO. \qquad ...(4)$$

Hydroxyl radicals may be formed from the peroxy radical by the following reaction:

$$C_4H_9OO \rightarrow \underset{\text{methyl ethyl keton}}{CH_3CH_2CCH_3} + OH \qquad ...(5)$$

The hydroxyl radical may react with additional *n*-butane

$$C_4H_{10} + OH \rightarrow C_4H_9 + H_2O \qquad ...(6)$$

to form more *n*-butyl radical. The *n*-butyl radical may participate in Reaction (4). Thus Reactions (4) through (6) constitute a chain in which hydroxyl radical is the chain-propagating free radical. The above reactions demonstrate some of the more important aspects of free radical reactions which are involving hydrocarbon in the atmosphere.

There is first of all a *primary photochemical reaction,* in this case the phtotodissociation of NO_2 to NO and atomic oxygen, Reaction (1). Next comes a *chain-initiating reaction,* in this case the reaction of atomic oxygen with *n*-butane, Reaction (3). The chain of Reactions (4) through (6) may continue until the chain-propagating free radical is destroyed in a *chain-terminating step.* Chain termination may take place in several ways, including reaction of two free radicals,

$$2OH \rightarrow H_2 + O \qquad ...(7)$$

adduct formation with nitric oxide or nitrogen dioxide,

$$OH + NO_2 + M \rightarrow HNO_3 + M \qquad ...(8)$$

(Here M refers to a third body which absorbs energy), or reaction of the radical with a solid particle surface. Gas phase reaction chains may have as many as one million steps. Furthermore, chain-branching reactions in which a free radical reacts with an excited molecule making it to produce two new free radicals may occur.

Free radicals are quite reactive: therefore, they generally have short lifetimes. Hydrocarbons may undergo heterogeneous reactions on particles in the atmosphere. Dusts composed of metal oxides or charcoal are having a catalytic effect upon the oxidation of organic compounds. Metal oxides may enter into photochemical reactions. For example, zinc oxide photosensitized by exposure to light has been found to promote oxidation of organic compounds

Phtochemical Smog

An oxidant in air pollution parlance of a substance in the atmosphere capable of oxidizing iodide ion to iodine (O). Sometimes substances other than iodide ion are used as reducing agents in the measurement of oxidants. A number of diffferent compounds—including O_3, NO_2, H_2O_2, organic peroxides, orgnic hydroperoxides, and peroxycyl nitrate—are oxidants in the atmosphere. Organic peroxides are having the general formula ROOR', and organic hydroperoxides are having the formula ROOH. Probably, the most notorious organic oxidant has been peroxyacetyl nitrate, PAN which has the formula as follows:

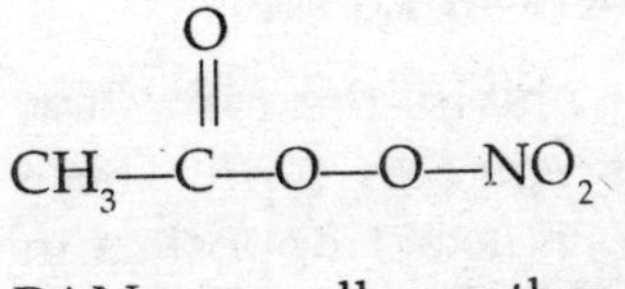

PAN, as well as other member of its homologous series of compounds, is produced photochemically in atmospheres having olefins and NO_x. It is damaging to plants, attacking younger leaves and bringng about "bronzing" and "glazing" of their surfaces. Exposure for several hours to an atmosphere containing PAN at a level of only 0.02 to 0.05 ppm will damge vegetation. The sulfhydryl group on proteins has been susceptible to damage by PAN. In reacting with sulfhydryl groups PAN may act as both an oxidizing agent and an acetylating agent.

Peroxybenzoyl nitrate, PBN, occurs in polluted atmospheres. It is a powerful eye irritant and lachrymator. It may be produced in the atmosphere by photochemical reactions which are involving aromatic hydrocarbons, nitrogen oxides and ozone.

In addition to PAN and PBN, some other specific organic oxidants which may be important in polluted atmospheres include peroxypropinyl nitrate (PPN) ; peracetic acid, $CH_3(CO)OOH$; acetylperoxide, $CH_3(CO)OO(CO)CH_3$; ethyl hydroperoxide, CH_3CH_2OOH; n-butylhydroperoxide, $CH_3CH_2CH_2COOH$; and *tert*-butylhydroperoxide, $(CH_3)_3$ COOH.

During the 1940s the Los Angeles area in California began experiencing a new and serious form of air pollution which was characterized by reduced visibility and eye irritation. The term "smog" has since been used to describe this kind of air pollution. The term is a misnomer for it was originally used to describe the unpleasant combination of smoke and fog laced with sulfur dioxide formerly very prevalent in London.

However, either smoke or fog nhibits the formation of photochemical smog of the Los Angeles type. These two types of smog are in many ways exact opposites in their chemical behaviour. The mixture of smoke, fog, and sulfur dioxide which afflicted London so badly for several centuries after the introduction of coal as a fuel is chemically a reducing mixture and is termed as a *reducing smog*.

The smog which permeates the Los Angeles area of number of days each year has a high concentration of oxidants and is an oxidizing mixture. It is therefore termed as an *oxidizing smog.* Of course, readily oxidized sulfur dioxide cannot last long in an atmosphere where oxidizing photochemical smog is present.

The formation of oxidants in the air—particularly ozone, but including PAN and other—has been indicative of smog formation. Serious levels of photochemical smog may be assumed to be present when the oxidant level exceeds 0.15 ppm for more than one hour. Ultraviolet light initiates smog formation and photochemical processes are involved.

Hydrocarbons must be present, and olefins are especially conductive to smog formation. Nitrogen oxides are required for the formation of smog, and a level of approximate 0.15 ppm NO_x is optimum. Apparently nitrogen oxides are involved in the primary photochemical processes leading to smog formation.

As it has been established that nitrogen oxides can inhibit atmospheric photooxidation of hydrocarbons, probably by queening reactions involving free radaicals, some concern exists that partial success in reducing nitrogen oxide emissions in some localities might actually increase the incidence of smog. Smoggy atmospheres show characteristic variations with time of day in levels of NO, NO_2, hydrocarbons, aldehydes, and oxidants.

The time-variations in levels of hydrocarbons, ozone, NO, and NO_2 are explained in a group of overall reactions first proposed by Friedlander and Seinfeld:

1. *Primary photochemical reaction:*

$$NO_2 + h\nu \rightarrow NO + O \qquad ...(1)$$

2. *Reactions involving oxygen species :*

$$O_2 + O + M \rightarrow O_3 + M \qquad ...(2)$$

(M is an energy-absorbing third body)

$$O_3 + NO \rightarrow NO_2 + O_2 \qquad ...(3)$$

3. Production of organic free radicals from hydrocarbons, RH :

$$O + RH \rightarrow R. + \text{other products} \qquad ...(4)$$

$$O_3 + RH \rightarrow R. \text{ and/or other products} \qquad ...(5)$$

R. is a free radical which may, or may not, contain oxygen.

4. Chain propagation, branching and termination:

$$NO + R. \rightarrow NO_2 + R \qquad ...(6)$$

In this case R contains oxygen and oxidizes NO. It is one of many chain propagation [reaction, some of which involve NO.]

$$NO_2 + R. \rightarrow \text{products (e.g., PAN)}$$

This is the most common chain terminating reaction, which occurs because NO_2 is a stable free radical present at high concentrations in a smoggy atmosphere. Reaction with NO may also occur. Chains may be terminated by reaction of two R. radicals, although this is uncommon because of their relatively low concentrations. Radical sorption on a particle surface is also possible and may contribute to particle growth. Many specific reactions are involved in the above overall scheme for the formation of photochemical smog. The formation of atomic oxygen by the primary photochemical reaction

$$NO_2 + hv \rightarrow NO + O \qquad ...(7)$$

leads to several reactions involving oxygen and nitrogen oxide species:

$$O + O_2 + M \rightarrow O_3 + M \qquad ...(8)$$

$$O + NO + M \rightarrow NO_2 + M \qquad ..(9)$$

$$O + NO_2 \rightarrow NO + O_2 \qquad ...(10)$$

$$O_3 + NO \rightarrow NO_2 + O_2 \quad ...(11)$$

$$O + NO_2 + M \rightarrow NO_3 + M \quad ...(12)$$

$$O_3 + NO_2 \rightarrow NO_3 + O_2 \quad ...(13)$$

There are a number of significant atmospheric reactions involving nitrogen oxides, water, nitrous acid, and nitiric acid:

$$NO_3 + NO_2 \rightarrow N_2O_5 \quad ...(14)$$

$$NO_3 + NO \rightarrow 2NO_2 \quad ...(15)$$

$$NO_3 + NO_2 \rightarrow N_2O_5 \quad ...(16)$$

[The reverse of Reaction (16) also occurs in a significant extent.]

$$N_2O_5 + HO \rightarrow 2HNO_3 \quad ...(17)$$

(Some of the nitiric acid produced in Reaction 17 may react with ammonia or particulate basic material to form particulate nitrates, which cause severe corrosion problems to, for example, electrical relay contacts.)

$$NO + HNO_3 \rightarrow HNO_2 + NO_2 \quad ...(18)$$

$$HNO_2 + HNO_3 \rightarrow H_2O + 2NO_2 \quad ...(19)$$

$$NO + NO_2 + H_2O \rightarrow 2HNO_2 \quad ...(20)$$

$$2HNO_2 \rightarrow NO + NO_2 + H_2O \quad ...(21)$$

Very reactive hydroxyl radicals, HO, can be formed by the reaction of excited atomic oxygen with water,

$$O + H_2O \rightarrow 2HO \quad ...(22)$$

photodissociation of hydrogen peroxide,

$$H_2O_2 + h\upsilon \rightarrow 2HO \quad ...(23)$$

or by the photolysis of nitrous acid,

$$HNO_2 + h\upsilon \rightarrow HO + NO \quad ...(24)$$

Among the inorganic species with which the hydroxyl radical reacts are oxides of nitrogen.

$$HO + NO_2 \rightarrow HNO_3 \qquad ...(25)$$

$$HO + NO + M \rightarrow HNO_2 + M \qquad ...(26)$$

and carbon monoxide,

$$CO + HO + O_2 \rightarrow CO_2 + HO_2 \qquad ...(27)$$

The last reaction is significant in that it may be responsible for the disappearance of some atmospehric CO and because it produces the hydroperoxy radical HO_2. One of the major inorganic reactions of the hydroperoxy radical is the oxidation of NO :

$$HO_2 + NO \rightarrow HO + NO_2 \qquad ...(28)$$

The kinetic rate constants for most of the above reactions are reasonably well known, enabling calculations of species concentrations to be expected under various conditions. These calculations, supported by laboratory studies of irradiated mixtures, are not able to explain the rapid transformation of NO and NO_2 that occurs in an atmosphere undergoing photochemical smog formation. In fact, except at very high CO concentration the concentration of NO_2 is predicted to remain at very low levels.

However, in the presence of reactive hydrocarbons, NO_2 accumulates very rapidly by a reaction process beginning with its photodissociation. It can be concluded, therefore, that the organic compounds form species which react with NO directly, rather than with NO_2.

Aliphatic hydrocarbons, RH, can react with O,

$$RH + O + O_2 \rightarrow ROO. + HO \qquad ...(29)$$

or HO radical,

$$RH + HO + O_2 \rightarrow ROO. + H_2O \qquad ..(30)$$

to produce relative oxygenated organic radicals, ROO. Olefins R = R' are much more reactive, undergoing overall reactions such as the following:

$$R = R' + O + 2O_2 \rightarrow ROO. + R''\overset{\overset{O}{||}}{C}OO. \qquad ...(31)$$

$$R = R' + O + 2O_2 \rightarrow ROO. + HO_2 + CO \qquad ...(32)$$

(In reactions such as 32 and 38 below where the HCO radical is formed, the HCO radical is believed to react with O_2 to form CO and H_2O radical.)

$$R = R' + HO = O \rightarrow ROO. + H\overset{\overset{O}{||}}{C}R'' \text{ (an aldehyde)}$$
..(33)

Aromatic hydrocarbons, H, may also react with O and HO :

$$ØH + O + O_2 \rightarrow ØOO. + H_2O \qquad ...(34)$$

$$ØH + HO + O_2 \rightarrow ØOO. + H_2O \qquad ...(35)$$

Aldehydes, $H\overset{\overset{O}{||}}{C}R''$, react with HO,

$$H\overset{\overset{O}{||}}{C}R'' + HO + O_2 \rightarrow R''\overset{\overset{O}{||}}{C}OO. + H_2O \qquad ...(36)$$

$$H_2C = O + HO + 3/2O_2 \rightarrow CO_2 + H_2O + H_2O \qquad ...(37)$$
(formaldehyde)

and undergo photochemical reactions:

$$H\overset{\overset{O}{\|}}{C}R'' + h\nu + 2O_2 \rightarrow R''OO. + CO + HO_2 \qquad ...(38)$$

$$H_2C = O + h\nu\ 2O_2 \rightarrow CO = 2HO_2 \qquad ...(39)$$

From the above reactions it is seen that hydrocarbons reacting with atomic oxygen or hydroxyl groups produce oxygen-containing free radical which are highly reactive. The reactions of these radicals with NO

$$ROO. + NO \rightarrow RO. + NO_2 \qquad ...(40)$$

$$R\overset{\overset{O}{\|}}{C}OO. + NO + O_2 \rightarrow ROO. + NO_2 + CO_2 \qquad ...(41)$$

are responsible for the rapid conversion of NO to NO_2. Peroxyacyl nitrates (PAN) are formed by an addition reaction with NO_2:

$$R\overset{\overset{O}{\|}}{C}OO. + NO_2 \downarrow R\overset{\overset{O}{\|}}{C}OONO_2 \qquad ...(42)$$

Alkyl nitrates and alkyl nitrites may be formed by the reaction of alkoxyl radicals (RO.) with nitrogen dioxide and nitric oxide respectively:

$$RO. + NO_2 \rightarrow RONO_2 \qquad ...(43)$$

$$RO. + NO \rightarrow RONO \qquad ...(44)$$

Addition reactions with NO_2 such as these are important in terminating the reaction chain involved in smog formation. Since NO_2 is involved both in the chain initiation step (Reaction 1) and the chain termination step, it is unlikely that moderate reductions in NO_x emissions alone will have much influence on curtailing smog formation.

The reaction of oxygen with alkoxyl radicals

$$RO. + O_2 \rightarrow HO_2 + H\overset{\overset{\large O}{\|}}{C}R'' \qquad ...(45)$$

produces hydroperoxy radical. Peroxy radicals are able to react with one another.

$$HO_2 + HO_2 \rightarrow H_2O_2 + O_2 \qquad ...(46)$$

$$HO_2 + ROO. \rightarrow RO. + HO + O_2 \qquad ..(47)$$

$$ROO. + ROO. \rightarrow 2RO_2 + O_2 \qquad ...(48)$$

to produce reactive hydrogen peroxide, alkoxyl radicals and hydroxyl radicals. Recently it is argued that electronically excited singlet oxygen designated here as O_2^* (g), may be an important oxidizing agent in the atmosphere, and contribute to the atmospheric oxidation of NO to NO_2. The abosrption of light by ground-state oxygen, O_2 (3Σg) takes place in the wavelength regiouns of 687 nm and 759 to 766 nm. The absorption is too weak to account for any appreciable level of singlet oxygen in the atmosphere. However, oxygen may be excited to the singlet state by energy transfer from another excited species.

Ultraviolet irradiation of mixtures of benzaldehyde and oxygen has produced singlet oxygen as evidenced by its emission of light of 1.27 micron wavelength. Although it has not been established that singlet oxygen participates significantly in atmospheric oxidation processes, it is known to react with olefins for forming peroxidic organic compounds. These compounds could then decompose to form radicals of the type RCO, thus starting the chain reaction process involved in smog formation. Aerosol particles are formed by the polymerization of the smaller molecules which are produced in the primary smog forming reactions. The chemical composition of the aerosol particles shows a high percentage of oxygen. Ether-soluble aerosols collected from the Los Angeles atmosphere show an empirical formula of approximately CH_2O. Therefore, the original hydrocarbons are strongly oxidized by reactions in the atmosphere.

London Smog

That there is a fundamental connection between air pollution and damge to health has been recognized for many years. Doctors, scientists and public health centers were altered to the problem when serious smog disasters occurred in the valley of the River Maas in Belgium in 1930 and in the city centre of Donora in the United States in 1948. Many people died as a result, and the coincidence in time of the atmospheric pollution and the resulting health damage highlighted the causes of the maladies and deaths. Considerably more action for the reduction of atmospheric pollution was promoted by the occurrence of fog in London. From December 5 to 9, 1952, many regions in Great britain were cloaked in fog. Particularly affected was the wide valley of the river Thames with London in its midst.

Within about 12 hours of the appearance of the heavy fog (which owing to a sharp inversion of weather conditions entrapped all the waste gases from the city), an extremely large number of people in London were smitten with breathing difficulties. This was soon followed by an unusually high number of deaths. The symptoms of the malady were severe coughing and expectoration, increased nasal secretions, sore throats and sudden vomitting. Generally speaking the troubles began without warming. Many of the most serious cases, mainly people already suffering from chronic bronchitis, asthma or pulmonary fibrosis, occurred on the third or fourth day of fog. The symptoms of patients seriously affected were shortness of breadth, cyanosis, fever, rattling in the throat and bronchospasms. A subsequent assessment of the mortality figures revealed that in 14 days of the smog disaster 4000 more people died altogether than was normal in the period.

From the very first day of the catastrophe, the number of deaths increased significantly; and the increase was greater in the center of London than in the outer ring of the city. A further 4000 people died in the following weeks and months. All age groups were affected, but those over 45 years of age and babies of up to 1 year most severely. In the age group from 1 to 14 years the death rate during the smog catastrophe was 1.3 times the previous rate, for over 45 year olds 2.8 times, and for babies 2.2 times.

During the smog disaster the concentration of sulfur dioxide in the atmosphere amounted on the average to 0.7 ppm, or about 6 times the normal level. The smoke concentration was about 5 times the normal. At that time it was still assumed that such a level was medically harmless. Subsequently, a number of other cases of smog occurred in London: January 3 to 6, 1956 (1000 deaths); December 2 to 5, 1957 (800); December 5 to 10, 1962 (700); and January 7 to 22, 1963 (700). It was only after these further catastrophes, in which thousands more victims perished, that atmospheric pollution over London was severely cut back by intensive counter-measures so that the danger of serious smog disasters in London now seems to have been averted. By a change in heating techiques and by insisting on the use of smokeless fuels, the hours of sunshine in winter have been raised in London by 70%.

Effects Phtochemical Oxidants on Plants

Injury to vegetation was one of the earliest observed results of photochemical air pollution. Available information suggests that ozone is the most toxic photochemical to plants, but continuing research is uncovering the importance of the PAN family. The effects of ozone on some particular plants are given in Table 2.

Table 2
Toxic Effects of Ozone on Plants

Plant	Ozone Level (ppm)	Exposure Time	Observed Effects
Radish	0.05	20 days (8-hr/day)	50% reduction in yield
Carnation	00.7	60 days	50% reduction in floral development
Tobacco	0.10	5-5 hours	50% reduction in pollen germination and pollen tube growth

Visible effects of ozone injure to plants are bleached or light flecks of stipples (small clusters of dead cells on the upper surface of leaves. Fully expanded, mature leaves are most easily damaged. Emergence tip burn, a disease of white pine, puzzled foresters and plant pathologists for years. It has been found that the presence of ozone correlates with the occurrence of this disease.

The other major secondary photochemical pollutant is peroxyacetylnitrate. Other common peroryacylnitrates are peroxypropionylnitrate (PPN), peroxybutyrylnitrate (PBN) and peroxyisobutyrylnitrate ($P_{iso}BN$). Preliminary studies show than PPN is several times as toxic to plants as PAN, while PBN and $P_{iso}BN$ are still more toxic than PPN.

Plant injury from PAN shows up as a glazing and bronzing of the lower leaf surfaces and a general attack on younger leaves. Eventually, leaf tissue dies. Exposures of a few hours to PAN concentrations of 0.02 to 0.05 ppm are sufficient to cause injury to vegetation. Five hours of exposure to a PAN level of 0.01 ppm will injure the most sensitive plants. It has been found especially toxic to citrus, forage and salad crops, and coniferous trees.

Ethylene (CH) is the only hydrocarbon known to have adverse effects on vegetation at ambient concentrations of 1 ppm or, less. Acetylence and propylene have shown some toxicity toward plants but only at concentrations of from 60-500 times as high as that of ethylene. The principal effect of ethylene upon plants is inhibition of growth. It also brings about color changes in leaves and death of flowering parts.

Effects of Photochemical Oxidants on Humans

Currently, there is no evidence to indicate that hydrocarbons, at present ambient air concentrations, exert any direct undesirable effects on humans. Experimental data obtained from research on humans and animals indicate that acyclic and alicyclic hydrocarbons produce undesirable effects only at concentrations hundreds to thousands of times higher than those now found in the atmosphere. No effects have been reported for levels lower than 500 ppm. Aromatic hydrocarbons are a greater threat than the acyclinc alicyclic. The vapours are much more irritating to the mucous membranes, and systemic injury can result from the inhalation of aromatic compound vapours. However, no effects have been reported at air concentrations lower than 25 ppm. Some effects of two common aromatics are given in Table 3. Photochemical oxidants enter the body as part of inhaled air and at sublethal concentrations alter, impair of otherwise interfere with normal respiratory processes. Some eye irritation is also characteristic of exposure to these compounds.

The following effects have been observed in humans exposed to ozone under experimental conditions: in general, no ill effects were noted at concentrations up to 0.2 ppm and a level of about 0.3 ppm appeared to be the threshold level at which nose and throat irritation began. Specifically, exposure to ozone concentrations of 1.0 to 3.0 ppm for a period of 2 hours produced extreme fatigue and lack of coordination in sensitive experimental subjects. Exposure to concentrations of about 9.0 ppm for similar time periods produced severe pulmonary edema in most subjects.

Table 3

Toxicity of two Aromatic Hydrocarbons

Compound	*Concentration (ppm)*	*Effect*
Benzene	100	Mucous membrane irritation.
C_6H_6	3,000	Endurable 1/2 hour.
	7,500	Dangerous after 1/2 hour
	20,000	Fetal after 5-10 minutes.
Toluene	200	Mild fatigue, weakness, and
C_7H_8		confusion after 8 hours.
	600	Loss of coordination, pupils
		dilated after 8 hours.

It appears that at the concentrations known to occur in ambient atmospheres, PAn (peroxyacetylnitrate) does not present any recognized health hazard. It has been linked to eye irritation, but it appears to play only a minor role. Another member of the family, peroxybenzoylnitrate, more readily brings about eye irritation.

Effects of Photochemical Oxidants on Materials

Some researchers feel that much of the degradation of materials now attributed to "weathering" is actually the result of attack by air pollutants, particularly photochemical oxidants. Most of the research in this area has dealt with one oxidant, ozone, and two types of materials, rubber and textiles. It is known that many organic polymers, including rubber and natural and synthetic textiles, are subject to chemical alteration upon exposure to very small amounts of ozone. Ambient air concentrations are frequently high enough to cause these reactions to take place. The susceptibility to attack increases with increasing numbers of carbon-to-carbon double bonds in the material. The double bond is the site of such attacks.

Two different effects are produced by these reactions, carbon-chain breaking and carbon-chain cross-linking. In the first, the long chains of carbon atoms that make up the polymer are broken, and the material becomes more fluid-like and loses tensile strength. The second effect results in the formation of new links between parallel carbon chains. This causes the material to become less elastic and more brittle. This type of activity takes place in ambient ozone levels much lower than those known to affect humans.

Rubber in a relaxed state can be exposed to ozone for long periods of time without forming any characteristic cracks. However, exposure to ambient concentrations of 0.01–0.02 ppm ozone will result in the formation of cracks if the rubber is under a strain of as little as 2 or 3%.

Apparently, rubber is protected by a coating consisting of a compound of rubber and ozone (ozonide) which forms on the surface and protects against further penetration and attack by ozone. This protective film cracks when the rubber is stretched and fresh rubber surfaces are exposed to the atmospheric ozone. The main use of rubber is in automobile tires and such rubber is exposed to atmospheric ozone while in a state of continual tensile stress (stretching and flexing).

Table 4
Effects of Ozone on Rubber

Concentration of Ozone (ppm)	*Time for First Crack to Appear (Minutes)*
0.02	65
0.26	5
0.45	3
20,000 (2%)	1 second, cracks formed instantly

In order to prevent ozone attack and subsequent cracking antiozonant additives are additives are added to the rubber. These prevent ozone attack but quite expensive, costing about 0.50 per tire. Table 4 shows the effects of ozone on unprotected rubber.

7 SYSTEM APPROACH TO ENVIRONMENTAL PROBLEMS

Man's environment has suffered severe degradation because of failure to evaluate problem areas on a total systems basis and because of unforeseen consequences. Rachel Carson called attention to instances of unforeseen results from the use of chemical pesticides. Biological concentration of DDT illustrates the interactions in the biological systems. The wide use of nondegradable detergents dramatically demonstrated channels of population and also resulted in objectionable foaming. The present handling of solid wastes in urban areas reveals many deficiencies in failing to consider the handling in the dwelling, collection and transportation and the disposal, recycling, or reclamation as a complete system. The disposal method, incineration, hog feeding, or land filling, may dictate the segregation and handling at the dwelling. Economic, legal and political factors, and public attitude may be decisive in selecting a method of disposal. The collection of all sewage at a single plant may make possible the construction of a more efficient treatment plant, but the increased absolute waste load, even after treatment at the high efficiency, may be more than the stream can assimilate. Discharge of the wastes at several points, even at lesser efficiencies in removal, might have been possible with no difficulties. Failure of highway planners to consider noise, air pollution, and social disruption from highway alignment has resulted in citizen action to block proposed "improvements." How can we avoid such pitfalls in environmental planning! The basic land use pattern which is adopted, or allowed to develop haphazardly, is a significant determinant for the future needs and problems of the area. The livability and character of development is related to the density and distribution of the population. Once such things as to type of dwelling (individual detached homes or high rise apartments), transportation system, allocation of land uses, etc. have been allowed to develop, major changes are difficult and expensive to accomplish. Environmental planning should be considered at the time of land use planning and development. The Department of Agriculture can supply data on soils which will reveal areas most suitable for parks, for residential lawns, or for absorbing the liquid effluent from septic tanks.

Geological surveys will show the suitability for foundations, underground water resource potential, and areas suited for landfall waste disposal. New subdivisions in urban areas seldom consider these factors and do not make adequate provision for solid waste and sewage disposal. In densely populated areas,sewage collection and disposal should be provided initially. Septic tanks sometimes fail after a few years of use by city dwellers accustomed to sewers. Solid waste handling in high rise apartment and office buildings requires as much care in engineering planning as climate control, structure, and other factors. Functional use of spaces inside and outside the dwelling unit for satisfaction of human needs, both individual and family, must be considered as an interrelated system. Spaces for individual privacy, family activities, and community social interaction are required. Stimulating variety is essential to human development and preservation of individuality. Frequently, in the past, aesthetics and environmental problems such as noise and vibration, traffic, air pollution, and waste disposal, have been neglected by industry. As a consequence, past practice has been to separate industrial areas by buffer zones. The result has been tedium and stress of long travel time to work and increased pollution from automobiles. The imaginative approach to a pleasing architectural design of modern industrial buildings provisions of landscaped open space, are the solution of commercial and industrial development in the residential area. Rigid performance standards are essential in such circumstances. Waste disposal facilities have been similarly shunned as neighbours. Landfills, incinerators,and sewage treatment facilities have given problems where operation was faulty. With attractive architectural treatment and proper operation, adequate facilities become more acceptable. Analysis of an environmental problem on a systems basis will result in a better solution of that problem.

However, a piecemeal attack on one problem after another will not necessarily result in a total environment which maximizes benefits to man's health. There is interaction between areas. A frequent illustration is that of an industrial hygiene problem. It may be solved by installing an exhaust system, only to create an air pollution problem. The air pollution problem is solved by installing a liquid scrubber, only to create a water pollution problem. The water pollution problem is solved by precipitating the pollutant. Now, what does one do with the solid waste? The land, water and air resources on earth are limited. As population increases,pollution problems are intensified. As the magnitude of pollution potential increases, higher and higher efficiencies in treatment are required to maintain the status quo. Soon 100% recovery of pollutants may become necessary. Under such circumstances, the prevention or elimination of waste becomes critical. Measures to reduce the amount of the waste are next in importance.

The irreducible wastes remaining would present no disposal problem if they could be recovered and recycled or put to a useful purpose as raw material or to reclaim land as in the case of strip mines. The interrelatedness of the various areas of the environment is again shown by the necessity to consider the water pollution potential of land reclamation projects using waste products. It has been shown that the evaluation and solution of environmental problems must consider biological interactions,human stress,land use patterns, and inter relatedness of environmental areas. In addition, political and economic influences and consequences may be critical and decisive factors. Many environmental problems, e.g., water resource development, land use, and waste (air, water, solid) management require "problem-shed," or regional, approaches to provide effective control. Political innovations may be required to effect regional environmental management. The importance of mathematics in all branches of science, technology, engineering, socio-economics, etc. is known to everyone of you.

Differential equation and integration are so common in all applied sciences that at every stage we have to turn to mathematics for solutions. The reason is that mathematics is an exact science. In contrast biology, for instance, is still regarded as an inexact science. This is because biology deals with life science and it started as purely descriptive science. It is inherently complex and enough discoveries have not been made to bring precision of the type mathematics has. Today in inexactness in biology is due to our lack of knowledge. Environment, it is defined as "the sum of all external conditions and influences affecting the life and development of organism. "This area is thus a conglomerate of all basic and applied sciences, engineering, socio-economics and law. Added to it is the fact that environment is for the living system, which naturally brings in a measure of inexactness in the areas of environment. The environment of region, country, state, district, city, town and village is influenced by several factors, because there is a high degree of location-specificity in environment, on account of an all round diversity starting, with ethnic considerations, and going over to geological, geographical, climatic, floristic, faunistic, social, economic and cultural considerations.India is a country where at this stage, centuries are co-existing. We have both near stone-age and near spaceage, cultures running parallel. We have human beings existing as part of nature in perfect harmony with it; taking from nature only what they require and nothing more and existing as a part of the eco-system, as one of the species.

At the other end is the socio-economic system of the human kind where he is a co-creator and in conflict with nature and is dominated by considerations of greed. There are all manner of integrating situations in between. We have thus problems that afflict both developing and developed countries. This is true of all phases of national developments. The situation in India is relevant to both developing as also to the developed world.Today, our thinking about environment comes from the West. Historically, our understanding of environment was holistic. The whole universe together with its creatures belong to the Lord (Nature). One can enjoy bounties or nature by giving up all greed. Implicit in the statement is that no creature is superior to others and human being should not have absolute power over nature, let no one species encroach on the rights and privileges of any other species. This element of sustainability is ingrained in this because the emphasis is to use nature without greed. Once an element of greed enters, exploitation starts and we cease to utilize nature for the good, the benefit and the welfare of our fellow human beings.

Sustainable development is therefore, a concept of good and sound economic growth, a growth that can be maintained indefinitely with no or minimal damage to the environment. Good environment will ultimately beget good economics. The next question arises as to what are the factors that can promote sustainable development. These are :

(a) Population stabilization and health care;

(b) Integrated land use planning and watershed management. Ensuring adequate use of cropland, woodland or grassland for food, fuel, timber and fodder;

(c) Revegetating marginal land and greening the uncultivated half;

(d) Water pollution control in rivers;

(e) Air pollution control in industrial pockets;

(f) Non-polluting renewable energy;

(g) Waste recycling and reuse;

(h) Conservation of biological diversity;

(i) Human settlements without congestion and slum improvement; and

(j) Environmental education and awareness.

The elements of the life support system, namely, air, water, land, flora and fauna and the incoming sunlight form an interacting system. Sustainability of development depends on such a complex system. Even problems like the carrying capacity of our land has number of dimensions like relationship between land area, topography, population growth, and production/productivity of food, fodder, fuel, timber, fertilizer, fibre, medicines, etc.

For answering such questions, it is necessary to develop a systems approach. Furthermore, Mathematical Modelling is one of the important methods to resolve such complex situations. The other approaches could be PERT (Programme Evaluation Review Techniques). CPM (Critical Path Method), OR (Operation Research) and so on. Systems Approaches have acquired a significant place in modern technological development because these help to look at the problems in totality particularly for situations where different components give an impression of not working in cohesion, thus enabling to deal with whole system involving many parameters; operational mechanisms together with available resources. All these aspects need to be taken up in an integrated approach. Systems Approach has a wide scope as it helps to resolve problems in optimal utilization of resources at minimal cost. For instance, water quality models can be prepared about surface water, estuaries, lakes and large rivers in selected stretches with use of water quality parameters. Equally important are models on ground water in which impact on ground water quality by solid waste disposal which leads to leaching of chemicals through rain water can be worked out with reference to rainfall pattern, soil porosity and height of ground water table. This would also include monitoring of parameters like chloride, (Phosphate, COD), BOD.Similarly, air quality modelling can be undertaken with reference to pollutant from growing industrialization. We need impact assessment of environment and therefore, short and long range transport of pollutants from industries in different terrains, meteorological conditions and in different industrial complexes. Validation of models would be needed by actual monitoring for operational and regulatory purposes as also air quality standards based on effects of pollutants on human beings, animals, plants and historical monuments.

Ecological risk assessment process

Ecological risk assessment is a process that evaluates the likelihood that adverse ecological effects may occur or are occurring as a result of exposure to one or more *stressors*. The term *stressor* describes any chemical, physical, or biological entity or condition that has the potential to induce adverse effects on any *ecological component*, which might be individuals, populations, communities, or ecosystems. The mandate of the EPA to regulate potentially harmful chemicals in the environment relates not only to humans but also to animals and plants; not only to individuals but also to populations; not only to single species but also to ecosystems. However, although the EPA has provided agency-wide guidance for performing human health risk assessments since 1986, attempts to develop rationale and consensus for a set of definitive guidelines for the protection of the environment have begun only recently.

Guidance for ecological assessment of hazardous waste sites was presented in 1989, and a proposed framework for general ecological risk assessment was published in 1992. Represents the major variables involved in ERA, and how they relate to one another. In it present state of development, this framework does not define regulatory requirements or even provided procedural guidance. Rather, it presents a suggested paradigm intended to foster a consistent approach to ecological risk assessment, identify key issues, and define terminology. Additional guidance documents will be published in the next few years. The proposed framework is consists of three major phases:

— problem formulation
— analysis
— risk characterization.

These main components are conceptually similar to those developed for human health risk assessment. Although this step is not part of the ERA framework, illustrates the importance of discussion between risk assessors and risk managers at both coping and risk characterization stages. The risk manager can advise the risk assessor on what information is necessary for informed decision making, whereas the risk assessor must provide the risk manager with a complete understanding of the assumptions, conclusion,s uncertainties, and limitations of the risk assessment. In addition, a role for verification and monitoring in the risk assessment process is indicated to validation of the process,as well as confirmation of specific predictions made on the basis of the risk assessment.

Monitoring of site conditions can aid in the verification process and may identify other areas or ecological receptors in need of risk assessment. Verification and monitoring actions thus permit evaluation of the accuracy of the risk assessment conclusions and the effectiveness and practicality of the selected remedy.

Problem formulation: The ecological risk assessment process beings when an adverse ecological effect is observed or when a stressor of potential concern is identified. The first phase of ecological risk assessment is *problem formulation*, a systematic planning and scoping process to establish the goals, breadth, and focus of the risk assessment. This process scientific considerations. The final product of the problem formulation process is a *conceptual model* that identifies the assessment endpoints, the data needed, and the analyses to be used. Careful planning is required in the process because of the complexity of ecological systems and their differential and inconstant "value" in terms of public opinion and policy. The risks of multiple ecological components may need to be considered. The objectives of the ecological assessment must be clearly defined and should reflect both primary ecological concerns and the information required for effective decision making. The initial steps in problem formulation are the identification and preliminary characterization of stressors, the ecosystem potentially at risk, and the ecological effects observed or expected.

Stressors: As mentioned previously, stressors may be chemical, physical, or biological in nature. For chemical stressors, a phased selection process can be used to identify COCs, there chemicals most likely to contribute significantly to risk. The process of identify COCs for ecological receptors accounts for chemical concentrations in relevant environmental media relative to background or base-line concentrations; their predicted mobility, persistence, transformation, and bioaccumulative potential in the environment; and their observed toxicological hazards to biota. A three-phases screening process can be used to identify COCs that may cause adverse impacts on environmental receptors. In Phase 1, detected compounds are screened with respect to detection limits, detraction frequency, and concentrations in blank samples.

Table 1: Relationship of human and ecological risk assessment paradigms

Human	Ecological
Problem formulation	
Hazard identification	Characterization of
Does-response assessment	ecological effects
Exposure assessment	Characterization of exposure
Risk characterization	Risk characterization

— Detection limits vary depending on the compound, the analytical instrument and method used, and the characteristics of the matrix. If the detection limits are acceptable for risk assessment purposes, target analytes not detected in any samples may be eliminated from further consideration.

— Infrequently detected chemicals may be artifacts of sampling, handling, or analytical errors. EPA guidance allows for the elimination of a chemical if it is detected in low concentrations in few samples of only one or two media, and there is no reason to believe that the chemical is site-related

— Blank samples provide a measure of accidental contamination of the sample. To prevent inclusion of nonsite-related chemicals in the risk assessment, concentrations of all chemicals detected in blanks are compared with concentration detected in site samples. A chemical can be eliminated from further consideration if its concentration in site samples is less than 10 times the blank concentration for common laboratory contaminates, or 5 times the blank concentration for other chemicals.

Table 2: Example of stressor characteristics

Characteristic	Descriptor
Type	Chemical, physical, or biological
Intensity	concentration or magnitude
Duration	Short- or long-term
Frequency	Single event, episodic, or continuous
Timing	Occurrence relative to biological event or cycles
Scale	Spatial extent and heterogeneity

The second phase involves comparing estimated concentrations in samples to media-specific back-ground data. If the mean concentration of a compound in a given medium is statistically higher than mean background levels, it will be considered a COC. As a result of this process, compounds likely to be site-related and detected at statistically significant concentrations will be selected as COCs. In the third phase of COC selection, concentrations of those contaminants identified in the first two steps are compared to reference toxicity data found in the literature to calculate. TQs. Those contaminants whose concentrations exceed their respective reference criteria are considered COCs.

Ecosystem potentially at risk: Knowledge of the nature of the ecosystem potentially at risk provides the context for the assessment and permits assessors to identify appropriate subjects for the ERA. The process of identifying the ecosystem at risk in influenced by how the risk assessment was initiated. If adverse effects were observed first, those effects can indicate areas and potential *receptors of concern,* species that play key roles in ecosystem structure and/or function. If a potential stressor was identified first, information on its temporal and spatial distribution patterns and expected effects may be useful in selecting appropriate ecological ROCs.

Potential ROCs include:

— *Species that are vital to the structure and function of the food web*: In general, loss of a few individuals of a species is unlikely to significantly diminish the viability of the population or disrupt the community or ecosystem of which it is a part. As a result, the fundamental unit for an ERA is generally the population rather than the individual, with the exception of threatened and endangered species.

— *Species that exhibit a marked toxicological sensitivity to the COCs*: Ecosystem function can be impaired if certain component species are particularly vulnerable to chemical exposure. Selection of ROCs is thus designed to ensure that benchmark criteria are protective of the most sensitive organisms actually present at the site.

— *Species that have unique life histories and /or feeding habits*: Significant impacts on such species might eliminate unique ecological niches, with unpredictable results on the ecosystem as a whole.

— *Species for which toxicological data are readily available in the scientific literature.* Although such species may not be "key" in the sense of occupying a critical ecological niche, the availability of data of their responses to COCs reduces uncertainty in the evaluation as a whole.

Ecological effects: Data on ecological effects can be obtained from a number of sources. In order to optimize the acquisition process, data are often collected in phases or tiers. The decision to terminate the ERA or to proceed to the next tier is based on objective decision points set at certain levels of exposure or effects. Tier 1 may use available data and relatively conservative assumptions, e.g., calculation of a TQ. If a TQ sophisticated and site-specific analyses. Tiers may also be designed to guide the investigation through a series of questions regarding ecological effects and potential causes. For example, Tier 1 might be a field survey to determine the reproductive success of predatory birds nesting at a Superfund site. If any adverse effects are evident, subsequent tiers would focus on identifying causes for the observed field effect. For example, eggs might be collected and analyzed for pesticides, heavy metals, or other chemicals of potential concern. The optimal approach to the assessment must be decided on a site-by-site basis, depending on the nature of the site, the information required to support the decision-making process, and the time and funds available.

Selection of endpoints: Once the ecosystem at risk, ecological effects, and characteristics of potential stressors are known, assessors can select ecologically based endpoints that are relevant to th decision making. Two types of endpoints are defined: assessment and measurement. An *assessment endpoint* are quantitative expressions of a measurable effect or characteristic designed to provide insight into potential impacts on the corresponding assessment endpoints. Both assessment and measurement endpoints may relate to ecological components at any level of ecological organization, from individual organisms to the entire ecosystem. Because *individual-level endpoints* refer to stressor effects on single organisms, they are not usually considered as assessment endpoints. *Population-level-endpoints* can be used to determine risks to an entire population of a species. *Community-level endpoints* can be used to evaluate whether the composition and abundance of species have been altered as a result of exposure to the stressors. *Ecosystems-level endpoints* are used to evaluate the impacts of stressors associated with various environments media on ecosystem function. At sites where contamination is relatively widespread, found in several habitat types, and/or may affect organisms in multiple trophic levels, consideration of endpoints at several levels is recommended to ensure that no significant component is overlooked. Endpoint evaluation is discussed further later in the chapter, and more information on endpoint types and selection is provided in EPA guidance.

Conceptual model development: Available information of the ecosystem, stressor characteristics, and ecological effects are combined to define possible *exposure scenarios*. Each scenario includes the ecosystem and its principal components, the stressor, its source, and how it may contact or affect the ecosystem, and the spatial and temporal scales of the interaction. Those scenarios considered most likely to contributed to risk incorporated into a *site conceptual model* that:

— describes the ecosystem potentially at risk
— Summarizes reasonable hypotheses of how the stressor(s), might affect ecological components illustrates the relationship between assessment and measurement endpoints.

Hypotheses considered most likely to contribute to risk are evaluated further in the second phase of the ERA, analysis.

Analysis: The *analysis* phase, based on the conceptual model developed in problem formulation, is a site characterisation process. In this phase, assessors identify ecosystems in which impacts caused by stressors may occur, the biota species that may be exposed, and the magnitude and spatial and temporal patterns of exposure.

Data on the effects of the stressors are summarized and related to the assessment endpoints selected in the problem formulation phase. The summary consists of two main interactive activities:

— characterizatio:. of exposure
— characterization of ecological effects.

Exposure characterization: The purpose of exposure characterization are to measure or predict the distribution of a stressor in space and time, and determine whether the stressor likely to contract ecological ROCs. For chemical stressors, a combination of modeling and monitoring data commonly determines the magnitude and extent of contamination. For nonchemical stressors, the pattern of change may depend on season or land-use practices. These stressors can be characterized by ground reconnaissance, aerial photography, or satellite imagery. The final product of the exposure characterization is the *exposure profile*, which quantifies the magnitudes of exposure through space and time. For chemical stressors, it is usually expressed in terms of intake per unit time. For nonchemical stressors, it may be expressed as the percentage of habitat removed per year or event.

Characterization of ecological effects. The goals of the second major component of the analysis phase, ecological effects characterization, are to identify and qualify the adverse effects potentially caused by the stressor(s), and evaluate cause-and-effect relationships to the extent possible. Measurement endpoints are evaluated in terms of their relationships to corresponding assessment endpoints selected during problems formulation. A variety of statistical and modeling methods may be used to quality the *stressor-response* relationship. If the measured response is not the assessment endpoint, a series of extrapolations is necessary to relate the exposure to the assessment endpoint. The product of the ecological effects characterization is a *stressor-response profile* that describes the stressor-response relationship and reviews available evidence of causality. When the stressor-response relationship is based on field surveys as opposed to controlled toxicity tests, the causal factors responsibility for observed effects may not be evident. Satisfaction of some or all of the following criteria supports a presumption of causality.

— *Strength*: Exposure to the stressor causes an observable reaction.

— *Consistency*: The association is repeatedly observed.

— *Specificity*: The effects is frequently associated with a particular stressor.

— *Temporality*: The effect only occurs after exposure to the stressor.

— *Presence of a biological gradient*: The degree of response depends on the degree of exposure.

— *Plausible mechanism of action*: The proposed interaction between stressor and ecological component is feasible.

— *Coherence*: The hypothesis is consistent with knowledge of ecology and ecotoxicology.
— *Experimental evidence*: The stressor-response relationship can be replicated experomentally.
— *Analogy:* Similar stressors cause similar response.

Risk characterization. In *risk characterization;* the final phase of both HRA and ERA, the stressor-response and exposure profiles developed in the analysis phase are integrated to evaluate the likelihood of adverse ecological effects associated with exposure to a stressor(s). The purpose of risk characterization it to provide as complete and accurate a picture a possible of the potential risks associated with contact or co-occurrence with stressors. Risks are described in terms of the assessment endpoint(s), and can be expressed qualitatively or quantitatively, depending on available data. The ecological significance of potential effects is discussed in terms of their types and severity, spatial and temporal patterns, and the likelihood of recovery.

Uncertainty: The conclusion s of nay ERA, and the decision and actions based on them, are only as sound as the assessment and measurement endpoints selected, the data collected and interpreted, the observations made and recorded, and the quality of the professional judgment applied. It is important to recognize that each stop of the ERA process has some associated degree of uncertainty. The risk assessor therefore provides the risk manager with a sense of the overall level of confidence that can be placed in the assessment by specifying and justifying all the assumptions made and parameter values used, analyzing the strengths and weaknesses of the analyses used, and estimating the uncertainties associated with their application to the subject site. To compensate for the substantial inherent uncertainty of ERA processes, risk assessors and managers generally choose conservative assumptions and parameters, and accept that results will err on the side of overprotectiveness. Although here are many similarities to the HRA process, several key issues are unique to ERA. These difference have substantial cost and time implications for all sites undergoing the ecological risk assessment process. In general, ERAs are more costly and time consuming than human assessments due to the complexity of the laboratory and technical analyses.

Population versus individual risk: The subjects of human health risk assessment are individuals of one species. *Home sapiens.* In contrast, the fundamental unit of ecological risk assessment is generally considered to be the population rather than the individual. Although the immediate effects of stressors in the environment are on individuals, the ecological significance of these effects is determined by the impact on the population of the affected species, and the ramifications of such impacts on the ecosystem as a whole. For example, the death of an organism is obviously catastrophic for the deceased individual but unlikely to impact the viability of the population of which it was member. On the other hand, a chemical that kills few organisms but reduces the reproductive success of a key species may have considerable ecological consequences.

Ecotoxicity testing: Although the primary interest in ERA is the population or some other ecological component greater that the individual, we are limited in evaluating effects on higher levels of ecological organization by a local of both assessment and measurement endpoints that can provide information relevant for decision making. As mentioned previously, the study of toxic perturbations of ecological systems is beyond the scope of classical toxicology for a number of reasons:

— Whereas classical human toxicology relies on extrapolation of toxicity data from a handful of mammalian species, ecotoxicology must rely on extrapolation from a few test species to an enormous number of species with a tremendous taxonomic range and spanning cast difference in size, physiology, and native habitat.

— The spatial and temporal heterogeneity of exposure and conditions in natural systems can cause large variations in the doses and responses observed.

— Organisms in the environment are rarely exposed to pure compounds alone but rather to complex products, and/or sequestered in the environment.

This lack of knowledge of environmental variables and limited ability to replicate them in the laboratory or control them in the field results in a high level of uncertainty in assessors' predictions of the effects of stressors on any given ecosystem component from laboratory toxicity tests. Despite this uncertainty, however, toxicity testing can provide valuable information as part of a tiered approach to ERA. The risk assessor must use professional judgment to decide which types of toxicity tests most effectively reduce the inherent extrapolation uncertainly, and at what point the uncertainty is reduced enough to allow a decision to be made regarding the potential or actual impact of a chemical stressor(s) on an ecosystem. Because the physical, chemical, and biological factors involved in the exposure and response of a receptor to a chemical are so complex and variable, uncertainty is best addressed by using standardized, site-appropriate, state-of-the -science producers whose results can be readily verified and compared with results reported elsewhere.

Types of toxicity tests: Standard test methods are published by the EPA, the American Society for Testing and Materials, and other governmental agencies an standard-setting organizations. Toxicity tests can provide data on the *acute* and *chronic* effects of chemicals on organisms. The choice of test species is a critical part of this process. Because different types of organisms vary widely in their responses to chemicals, toxicity tests should use species representative of the ecosystem under study. An appropriate test auspices is one that is:

— easily obtainable and easily maintained in the laboratory, with well-defined growth patterns and nutritional requirements

— of economic, recreation, and/or ecological importance, or similar to such species

— widely distributed in the environment, allowing comparison of data obtained from different areas

— highly sensitive to the COCs.

In some cases, tests of species that live at a site may be appropriate. However, most toxicity tests are conducted under controlled conditions in laboratories with samples of contaminated soil, water, or sediment collected from the site being assessed. The use of site media accounts for the site-specific characteristics that constitute the habitat and control chemical bio-availability. *In site* toxicity tests, conducted with either laboratory or resident species at the site itself, can provide information or organism response to actual field conditions. Common endpoints examined in acute and chronic toxicity tests are percent response in a group of test organisms exposed to site media, and concentration-response relationships for tests run at several different proportions of site media. The responses observed may be lethal or sublethal. The statistic usually calculated from acute lethality studies is the LC_{50} or LD_{50}. Where sublethal endpoints are observed the EC_{50} or ED_{50} is calculated. Chronic test results ar usually expressed as estimates of concentrations below which no adverse effects are expected. The *maximum acceptable toxicant concentration* is presented as two test concentrations: the *no-observed-effects- concentration*, defined as the highest test concentration that caused no significant toxic effects, and the *lowest-observed-effect-concentration,* defined as the lowest test concentration at which any adverse two value is an estimate of the threshold for chronic toxicity.

Future directions in ecotoxicity testing: It is widely agreed that ecosystem-level analyses are most relevant for ERAs. An active area of current research is therefore identifying and improving methods of extrapolating results observed in single organisms to the population level. Many researchers use population models for this purpose. However, extrapolation of the endpoints examined in single-species studies to multiple species at the community or ecosystem level is much more difficult. One strategy being developed to examine endpoints at the level of the ecosystem is toxicity testing with multiple species. Mulitspecies test systems include model ecosystems such as *macrocosms, mesocosms*, and *field tests*. Reviews of these and other methods cna be found in EPA and Suter. Although such tests are promising, the lack of well-defined ecosystems-level assessment endpoints presently limits the applicability of their results to ERAs.

Sediments: As part of PL 92-500, the EPA was mandated by Congress to develop federal water quality criteria. The federal criteria serve as a basis for each state's own water quality standards. Based on a broad range of toxicological studies using a wide range of water column organisms and life stages, these criteria were intended to endure protection of 95% of all aquatic organisms from adverse effects. However, significant environmental damage has occurred even in areas where WQC are met. In such cases, the primary source of exposure is *sediment*.

Aquatic sediment can be defined as a collection of debris-coarse to fine particles of mineral and organic matter deposited at the bottom of water bodies. Fifty to 90% of the volume is "interstitial" or "pore" water, depending on depth and degree of compaction. Organic matter content typically ranges from 0 to 10%. Sediments are an important component of aquatic ecosystems because of the habitat they provide for *benthic* organisms, including commercially important organisms such as shrimp, crayfish, lobster, crab, mussels, and clams, as well as species important in the environmental food web, such as many species of worms, amphipods, bivalves, and insects. Because of the tendency of both organic and inorganic molecules to adsorb to particles, sediments are also repositories for a wide variety of potentially toxic chemicals, both naturally occurring and anthropogenic. Chemicals undetected or found in only trace amounts in the water column can accumulate to high levels in sediments. These chemicals can enter the water from point sources , nonpoint sources, and from spills, leaks, or dumping of wastes. Common sediment contaminates include;

— organic chemicals such as halogenated hydrocarbons

— PAHs such as benzo[a]pyrene

— solvents, as well as metals

Although this "sink" effect serves to isolate chemicals from aquatic organisms, the sediment can also serve as a reservoir, slowly releasing chemicals from sediment to pore water to organisms. Thus, for the large number of commercial and food-chain organisms that spend a portion of their life cycles close to the sediment, sediment provides chemical exposure as well as nurture. In fact, direct transfer of chemicals from sediments to organism is now considered to be a major route of exposure for many benthic species. Further, consumption of benthic organisms by higher tropic level animals, such as fist-eating birds and humans, can be a major pathway of exposure to chemicals otherwise isolated in the sediment sink. Water quality criteria may not be protective of benthic species because they are principally based on studies with water column organisms that live and feed in the overlying waters as opposed to benthic organisms that live and feed in sediments. As a result, attention has shifted form point to nonpoint sources of pollution, and protection of sediment quality has emerged as a necessary extension of water quality protection. To evaluate the threat posed by contaminated sediments, several questions must be answered:

— What concentrations of contaminants cause adverse effects?

— What proportions of adsorbed contaminants are bioavailable?

— What are the factors that mediate the toxicity and bioavailability of contaminants in sediment?

Development of sediment quality criteria. Several environmental states-the Clean Water Act; the Toxic Substances Control Act; the Federal Insecticide, Fungicide, and Rodenticde Act; as well as RCRA and CERCLA-give EPE the authority to address these questions and develop methods for management of contaminated sediments. The agency is authorized under Section 304 of the Clean Water Act to develop and impement sediment quality criteria. Because the extent and significance of the problem are not well defined, and scientific approaches to the assessment and remediation of sediments are relatively new and unproven, this is a difficult charge. The EPA has conducted an analysis of 11 of the currently availabe methods for sediment quality assessment. These apporaches may be divided into two general types:

— approaches that provide criteria for individual chemicals
— approaches that consider the effects of the mixtures of chemicals that are usually encountered in the environment.

Examples of each are briefly described. Additional information on these and other approaches can be found in excellent reviews by Adamsetal and Chapman. The chemical-specific approaches yeild numerical criteria representing "safe" levels of indivdual chemicals, whereas chemical mixture approaches are used to identify sediments that cause adverse biological effects without specifying or quantifying the contributions of individual chemicals to these effects. Both approaches have strenghs and weaknesses, but neither is adequate for development of generally applicable sediment criteria. Although the chemicalspecific approaches are generic and easy to apply, they do not account for the effects of chemical interactions within the complex chemical mixtures found in most sediments. The EPA has proposed the equilibrium partitioning approach for derivation of chemical-specific SQO for nonionic organic chemicals. This method calculates an organic carbon-normalized SQO from WQC using the organic carbon: water partition coefficient Koc. It allows for differences in chemical bioavailability at different sites by multiplying the generic SQCoc by the site-specific fraction of organic carbon. Because organic carbon does not control the adsorption of metals or ionic organic chemicals to sediments, this approach is only applicable to nonionic organics for which WQC have been developed. Efforts to develop other normalization methids for these other chemical groups are currently in progress. The interpretive difficuly posed by the presence of multiple contaminants is avoided in the chemical mixture approaches, because they are based on observed effects on organisms rather than chemical concentrations.

Additional efforts needed: Greater understanding of the role of sediments as a reservoir of potentially toxic chemicals is necessary to protect public health and the environment. Unfortunately, development of sediment quality criteria is considerably more difficult than development of water quality criteria because of the many site-specific factors that influence the toxicity of chemicals in sediment but not in water. These factors derive from the nature of sediment. Unlike the overlying water, sediment is relatively structured and immobile, and consists of a somplex array of particles of different sizes and chemical characteristics. These characteristics influence the degree and affinity of chemical adsorption, and hence bioavailability. Adsorbed contaminants do not mix freely in sediment as they would in water, but depend on sediment as they would in water, but depend on sediment deposition and erosion patterns for movement. Thus, the concentration and distribution of sediment-associated chemmicals can be highly heterogeneous in both space and time. Because no existing assessment methodologies have undergone extensive review or field validation, the few criteria presently available cannot be regarded as more than preliminary. Immediate efforts are needed to:

— identify the extent of sediment contamination and prioritize areas of greatest potential concern
— incorporate existing approaches into a flexible tiered approach
— expand, improve, and vaildate existing assessment methodologies
— develop new approaches as conceptual and technological advances allow.

The purpose of ERA and the legislation that mandates it is to protect the 'environment' from physical, chemical, and biological "stressors" of all descriptions. This is a necessary goal, but it is also substantially beyond researches' current capabilities. This is partially because funds are lacking to support and skilled personnel to perform research and testing on the tens of thousands of extant chemicals and the physical changes occurring on a global scale. The other reason is the absence as yet of an intellectual framework to guide such efforts. This framework must be grounded in a new, multidisciplinary science melding concepts and approaches of ecology, toxicology, chemistry, physics, engineering, and other sciences. Development of an ecosystem-level perspective in ecotoxicology and ecological risk assessment is considered to be the first priority for advancing researchers' understanding of the impacts of anthropogenic stressors on ecosystems. The major challenges facing risk assessors are to define what should or must be preserved in the environment; to identify explicit, rational assessment endpoints reflective of these values; to develop corresponding measurement endpoint for quantitative evaluation of their responses to contact with stressors; and to develop accurate and costeffective methods of stressor evaluation that will reduce the uncertainty involved in extrapolation of results from laboratory to natural conditions or from one site to another. Because of the enormous complexity of the environment, its potential stressors, and their interactions, it is unlikely that numerical criteria can be developed to ensure environmental health. Even if they could, chemicals are not the only or even the most serious anthropogenic threat to the environment.

The risk assessment process synthesizes available data on exposure of specified receptors and the toxicity of potential chemicals of concern to estimate the associated risk to human health. The characterization of potential adverse human health effects resulting from exposure to chemicals involves five basic steps:

1. *Data collection:* All available information pertaining to a site must be carefully reviewed to determine basic site characteristics, identify potential exposure pathways and points, and determine data needs for risk assessment. Necessary information includes:
   - — physical characteristics of the site
   - — chemical and processes historically used at the site
   - — levels of contaminants present in various environmental media
   - — potentially exposed populations
   - — reliable site-specific data for various human activities and behavior patterns.
2. *Data evaluation:* Available site data are reviewed to characterize the site, define the nature and magnitude of chemical releases environmental media (soil, air, and wter), and identify potential site-related COCs, exposure media, and receptor populations. Key questions are identified and plans for addressing them are made at this stage.
3. *Exposure assessment:* Assessment of the degree of exposure involves estimation of the amount, frequency, duration, and routes of human exposure to site-related chemicals. The exposure assessment should consider both current and likely future site uses. Identifying exposure routes (such as inhalation and ingestion) and receptors (i.e., the person(s) who could come in contract with a chemical)are crucial to determining the validity of an exposure pathway. After potentially complete exposure pathways are identified, exposure point concentrations and chemical intakes by exposed humans are calculated.
4. *Toxicity assessment:* Assessing the degree of toxicity involves review of available information to identify

the nature and degree of toxicity of the COCs and to characterize the does-response relationship (the relationship between magnitude of exposure and magnitude of adverse health effects).

5. *Risk characterization:* Characterizing risk to the identified receptors involves combining exposure and toxicity information to determine the degree of potential risks at a site, and to estimate what residual levels of chemical may *not* pose unacceptable risks to receptors. This step also includes an analysis of the uncertainties associated with the risk estimates to put them and potential target criteria into appropriate perspective for decision making.

This sequence of events applies to human health risk assessments for both carcinogenic and noncarcinogenic chemicals. Therefore, the risk assessment is frequently used s basis for regulatory policy decisions because a variety of chemicals can be evaluated.

The information generated in the risk assessment can be used to:

— determine whether remedial action (cleanup)is necessary from a human health standpoint (i.e., if risks exceed some preestablished criterion)

— identify the contaminated media and specific areas that many require remediation

— provide a basis for establishing cleanup levels that must be achieved in order to adequately protect human health, if remediation is required.

A characterization of the physical setting should summarize information on climate, hydrogeology, geologic setting, and soil type, as well as the location and description of surface water and groundwater bodies on or near the site. An understanding of historical operations at the site can help guide the site characterization process by suggesting what chemicals might be found and in what form they might exist. Identification of chemicals present on-site that may also be present in the background (i.e., unrelated to operations on-site)is an important part of the site characterization process. Because the risk assessment typically quantifies health effects from potential exposures on-site chemicals exceeding background concentrations, careful evaluation of background concentrations can justify the exclusion of certain chemicals from the quantitative component of the risk assessment process. this is especially true for metals, which an be toxic under certain conditions but are natural components of the earth's crust (e.g., arsenic, cadmium, lead, and mercury). The widespread use and disposal of organochlorine pesticide and polychlorinated biphenyls (PCBs) have resulted in their detection in virtually all media (including human tissue worldwide). Other ubiquitous anthropogenic compounds, including polycyclic aromatic hydrocarbons (PAHs; e.g., benzo [a]pyrene) and chlorinated dibenzoxins and -furans, derive from combustion processes.

A description of potentially exposed populations should include information on sensitive subgroups (e.g., children, pregnant women, and the elderly) as well as activity patterns that could affect exposures. Such information can be obtained from local agencies and site-specific survey results. This section of the assessment should also include information on individuals who could be exposed to site-related contaminants even if they do not live on or near site. The risk assessment identifies potentially exposed populations associated with the following land-use categories:

— residential
— commercial/industrial

— recreational
— trespasser-a term used in a nonlegal sense, referring to episodic exposures at a site.

To accurately define current land uses, assessors often must gather information on activity patterns (e.g., determining the amount of the time residents, workers, or trespassers spend in potentially contaminated portions of the site, and assessing how activities and exposures can change with the seasons). Information on the proximity of individuals to potentially contaminated areas can be obtained from zoning maps, census data, and population surveys conducted on or near the site. Although future land use of the site is typically characterized by reviewing zoning plans, preparing census projections, and making reasonable assumptions on how long residents have lived on or near the site can site. For example, if the site is in the middle of an industrial area, assessors can judge how likely it is to be developed into a single-family residential area. The collection of site-specific data enables assessors to more accurately estimate risk in its geographical context and precludes the need to use default assumptions from agency guidelines. For instance, if warranted, assessors could conduct a survey of the people living on a site or who use a site to reduce the uncertainties of assuming default values for many exposure factors.

The goal of such an exposure assessment survey would be to obtain site-specific information on behavioral patterns and frequencies so incorporated in the risk assessment. In addition, this survey would permit the risk assessment delineate the factors that contribute most to exposure. Conducting a site-specific exposure survey can be a relatively inexpensive means of achieving more realistic risk estimates.

8 Future Prospects of Sustainability

Development is about improving the well-being of people. Raising living standards and improving education, health, and equality of opportunity are all essential components, of economic development. Ensuring political and civil rights is a broader development goal. Economic growth is an essential means for enabling development, but in itself it is a highly imperfected proxy for progress. The first step in improving social choices is to measure progress correctly.

It has long been recognized that measures of, for example, educational opportunity, infant mortality, and nutritional status are essential complements of GDP or GNP. Some have even tried to merge these indices to capture progress in development. The human development index constructed by the United Nations Development Programme (UNDP) is such an effort.

The fact that environmental damage hurts people, both today and in the future-provides additional grounds for rethinking our measurement of progress. Indeed it raises special concerns, for unlike education, health, nutrition, and life expectancy, which tend to be improved by economic growth, the environment is sometimes damaged by that growth. Furthermore, the people suffering from the damage may be different from those enjoying the benefits of growth. They may, for example, be today's poor, or they may be future. Generations who inherit a degraded environment. For these reasons it is essential to assess the costs to human welfare of environmental damage-a central theme of this Report-and to take account of the distributional impacts of policies, particularly for the poor. Sustainable development is development that lasts. A specific concern is that those who enjoy the fruits of economic development today may be making future generations worse off by excessively degrading the earth's resources and polluting the earth's environment..

The general principle of sustainable development adopted by the World Commission of environment and Development—that current generations should "meet their needs without compromising the ability of future generations to meet their own needs"—has become widely accepted and is strongly supported in this Report.

Turning the concept of sustainability into policy raises fundamental questions about how to assess the well-being of present and future generations. What should we leave to our children and grand children to maximize the chances that they will be no worse off than ourselves? The issue is the more complicated because our children do not just inherit our pollution and resource depletion but also enjoy the fruits of our labour in the form of education, skills, and knowledge (human capital), as well as physical capital. They may also benefit from investments in natural resources—improvement in soil fertility and reforestation, for example.

Thus, in considering what we pass on to future generations, we must take account of the full range of physical, human, and natural capital that will determine their welfare and their bequests to their successors. Intergenerational choices of this kind are reflected in the discount rate used to assess investments.

The discount rate is the mechanism through which present and future costs and benefits are compared. The lower the discount rate, the more it is worth investing today to make future gains. It is sometimes claimed that a lower discount rate even a zero discount rate-should be used in order to give appropriate weight to the long-term consequences of environmental change. This argument is erroneous. Provided that the environmental effects of projects are fully taken into account—which they often are not—it is always best to choose the investments which generate the highest net rate of return.

Encouraging investments that yield a lower net rate of return is wasteful; it implies a loss of welfare and of income that might have been devoted to environmental objectives. The term "sustainable development" was brought into common use by the World Commission on Environment and Development in its seminal 1987 report Our Common Future. The idea of sustaining the earth has proved a powerful metaphor in raising public awareness and focusing on the need for better environmental stewardship. The Brundtland Commission's definition of the term—meeting the needs of the present generation without compromising the needs of future generation"—is strongly endorsed by this Report.

We also believe, with the Brundtland Commission, that meeting the needs of the poor in this generation is an essential aspect of sustainably meeting the needs of subsequent generations. There is no difference between the goals of development policy and appropriate environmental protection. Both must be designed to improve welfare. Making the concept of sustainability precise, however, has proved difficult. It is not plausible to argue that all natural resources should be preserved. Successful development will inevitably involve some amount of land clearing. Oil drilling, river damming, and swamp draining. Some have argued that natural capital should be preserved in some aggregate sense, with losses in one area replenished elsewhere. This approach has helpfully focused attention on the need to estimate the value of environmental resources and on the importance of protecting certain essential ecological systems. This Report supports efforts to assess values but goes further. Societies may choose to accumulate human capital (through education and technological advance) or manmade physical capital in exchange, for example, for running down their mineral reserves or converting one form of land use to another. What matters is that the overall productivity of the accumulated capital-including its impact on the human health and aesthetic pleasure, as well as on incomes more than compensates for any loss from depletion of natural capital.

In the past benefits from human activity have often been exaggerated, and the costs of environmental loss have been ignored. These costs must be built into decisionmaking, and all short- and long-term impacts must be carefully explored. This cannot be done without taking account of the uncertainties and irreversibilities associated with some environmental process, recognizing some impacts occur far into the future. Not all environmental resources can or should be assigned monetary values, but tradeoffs should be made as explicit as possible.

It is sometimes argued that the benefits from human investment are temporary, while the benefits of an undisturbed environment last forever. This has prompted some to advocate using a lower discount rate in project analysis. But this may lead to more damage (through encouraging investment) rather than less. The answer lies not in artificially lowered discount rates but in ensuring that the benefits from an expanding economy are reinvested. Basing developmental and environmental policies on a comparison of benefits and costs and on careful macroeconomic analysis will strengthen environmental protection and lead to rising and sustainable levels of welfare.

When this Report uses "sustainable development" and "environmentally responsible development," it refers to this narrower definition. Addressing environmental problems requires not that discount rates be artificially lowered but rather that the value of the environment be factored into decision making.

Values that are difficult to measure are often implicit in decision making, but the tradeoffs are not well thought through. There is a clear need to make such costs and benefits as explicit as possible so as to better inform policymakers and citizens. This does not imply that it is possibly, or even desirable, to put monetary values on all types of environmental resources. But it is desirable to know how much environmental quality is being given up in the name of development, and how much development is being given up in the name of environmental protection.

This Report argues that too much environmental quality is now being given up. There is, however, a danger that too much income growth may be given up in the future because of failure to clarify and minimize tradeoffs and to take advantage of policies that are good for both economic development and the environment. To clarify these tradeoffs at the national level, efforts are under way in a -lumber of countries to amend the national accounts. Such exercises can be valuable for two reasons. First, they can help indicate how growth of GDP may bring with it environmental costs for today's citizens. For example, the costs of pollution to health and productivity should be taken into account in the same way that other measures of welfare need to be considered. Second, it can held give a more realistic measure of the capacity of an economy to produce. To this end, investment has to be adjusted to take account of depreciation of physical and natural capital. But the accumulation of human capital and the benefits of technical change must also be taken into account to provide an overall picture of an economy's productive capacity.A number of approaches to measuring environmental costs have been tried in different countries.

A recent pilot study of mexico's national accounts indicates the potential magnitudes of the adjustments required. When an adjustment was made for the depletion of oil, forests, and groundwater, Mexico's net national product was almost 7 per cent lower. A further adjustment for the costs of avoiding environmental degradation, particularly air and water pollution and soil erosion, brought the national product down another 7 per cent. These estimates are preliminary and are only intended to illustrate a methodology.

Of more value than these aggregates numbers are sectorial calculations. In the livestock sector, for example, adjustments for the costs of soil erosion sharply reduced the sectors' net value added. These calculations in themselves give no indication to policymakers as to whether Mexico's use of natural capital has been in the country's best interest, but they can be useful in reminding policymakers of potential tradeoffs and can assist in setting sectorial priorities. The adverse impact of economic growth on environmental degradation can be greatly reduced. Poor management of natural resources is already constraining development in some areas, and the growing scale of economic activity will pose serious challenges for environmental management. But rising incomes combined with sound environmental policies and institutions can form the basis for tackling both environmental and development problems. The key to growing sustainably is not to produce less but to produce differently.

In some situations, such as protection of forests or control of emissions, good environmental policies may cause short-term growth to fall, even as welfare may rise. In other cases—for example, improved soil conservation practices or investments in water supply—the effect on output and incomes is likely to be positive.

In still other areas the impacts are unclear. What is clear, however, is that failure to address environmental challenges will reduce the capacity for long-term development. All economic activity involves transforming the natural world. Why does economic activity sometimes lead to excessive environmental degradation? One reason is that many natural resources are shared and the true value of many environmental goods and services is not paid for by those who use them.

Some natural resources are shared because there is no mechanism for enforcing property rights, as with frontier land, and others are shared because, as with the atmosphere, property rights are impossible to enforce. Unless an explicit agreement among users emerges, shared resources will be degraded over time, particularly as the scale of population and economic activity increases.

In some cases government policies that subsidize environmental degradation can induce more damage than might otherwise occurred. In other cases the poor, with few assets on which to draw, may have no choice but to excessively degrade natural resources.

The most pressing environmental problems are associated with resources that are regenerative but are undervalued and are therefore in danger of exhaustion. Air and water are renewable resources, but they have a finite capacity to assimilate emissions and wastes. If pollution exceeds this capacity, ecosystems can deteriorate rapidly. When fisheries or forests are excessively depleted to meet human needs, critical thresholds maybe passed, resulting in the loss of ecosystems and species. Shortages of non renewable resources, such as metals, minerals, and energy, the possible exhaustion of which preoccupied early environmental debate, are of less concern.

The evidence suggests that when the true value of such non renewable resources is reflected in the market place, there is no sign of excessive scarcity.Water provides an example of an undervalued renewable resource that is showing signs of shortage. By the end of the 1990s six East African countries and all the NorthAfrican countries will have annual renewable water supplies below the level at which societies generally experience water shortage.

In China fifty cities face acute water shortages as groundwater levels drop I to 2 meters a year. In Mexico City groundwater is being pumped at rates 40 per cent faster than natural recharge. These shortages emerge when water is lost or wasted because its true scarcity value is not recognized. In such cities as Cairo, Jakarta, Lima, Maniala, and Mexico City more than half of urban water supplies cannot be accounted for. In many countries scare water is used for how-value agricultural corps, and farmers pay nothing for the water they use. The misuse of water in the Aral Sea in Central Asia is an extreme example of failure to recognize the value of a natural resource.

Assessment of whether the regenerative capacity of a natural resource has been exceeded is complicated by uncertainty about the effect of economic activity on the environment. In the cases of soil erosion, atmospheric pollution, and loss of biodiversity, there is often substantial scientific uncertainty about the extent of environmental degradation. Controversy also surrounds the consequences of degradation. What are the health effects of certain pollutants? What will climate change do to the ecosystem? Can tropical, forests be regenerated? The solutions are likewise often unclear. How quickly can be atmosphere restore itself? When will certain cleaner technologies become available and cost-effective? Uncertainty is an inherent part of environmental problems. To reduce it, decisionmakers need better information about environmental processes and social preferences. The view that greater economic activity inevitably hurts the environment is based on static assumptions about technology, tastes, and environmental investments.

According to this view, as populations and incomes rise, a growing economy will require more inputs (thus depleting the earth's "sources") and will produce more emissions and wastes (overburdening the earth's "sinks"). As the scale of economic activity increases, the earth's "carrying capacity" will be exceeded. In reality, of course, the relationships between inputs and outputs and the overall effects of economic activity on the environment are continually changing.

The scale of the economy is only one of the factors that will determine environmental quality. The key question is whether the factors that tend to reduce environmental damage per unit of activity can more than compensate for any negative consequences of the overall growth in scale. Factors that can play a particularly important role are:

- *Structure:* the goods and services produced in the economy
- *Efficiency:* inputs used per unit of output in the economy
- *Substitution:* the ability to substitute away from resources that are becoming scarce.
- *Clean technologies and management practices:* the ability to reduce environmental damage per unit of input or output.

Economic policies, environmental policies, and environmental investments all have a role in ensuring that individual behaviour takes account of the true value of environmental resources.

Economic policies affect the scale, composition, and efficiency of production, which can result in positive or negative effects on the environment. Efficiency gains from economic policies will often reduce the demand for natural resource inputs. Environmental policies can reinforce efficiency in resource use and provide incentives for adopting less-damaging technologies and practices. The investments that are induced by environmental policies will change the way in which goods and services are produced and may result in lower output but will also generate benefits than can increase human welfare.

As incomes rise, the demand for improvements in environmental quality will increase, as will the resources available for investment. Without incentives to use scarce resources sparingly, the pressure to reduce environmental damage will be weaker, and the adverse effects of economic growth are likely to dominate. But where the scarcity of natural resources is accurately reflected in decisions about their use, the positive forces of substitution, efficiency gains, innovation, and structural change will be powerful. In industrial countries these positive forces contributed significantly to improving environmental quality while maintaining economic growth. The Environmental problems facing poor countries differ from those facing the better off. In some cases environmental quality improves as income rises. This is because increased income allows societies to provide public goods such as sanitation services and because once individuals no longer have to worry about day-to-day survival, they can afford profitable investments in conservation.

Some problems are observed to get worse as incomes rise. Carbon dioxide emissions and municipal wastes are indicators of environmental stress that appear to keep rising with income. But this is because no incentives yet exist to change behaviour.

The costs of abatement in these cases are relatively high, and the benefits of change behaviour are perceived to below-partly because (in the case of carbon dioxide) they would accrue mainly to other countries. When societies have decided to enforce a change in behaviour-through regulations, charges, or other means environmental quality has improved. Progress in reducing water pollution and emissions of particulates, lead, and sulphur dioxide are examples of how higher income countries have been able to break the link between growth and environmental degradation. This is not easy it requires strong institutions and effective policies-but it can be done. It explains why so many environmental indicators show an initial deterioration followed by an improvement.

As incomes grow, the ability and the willingness to invest in a better environment rise. Past patterns of environmental degradation are not inevitable. Individual countries can choose policies that lead to much better (or worse) environmental conditions than those in other countries at similar income levels. In addition, technological change, coupled with improved understanding of the links between economic activity and environmental damage, is enabling countries to grow more rapidly with less environmental impact than was possible earlier. At any given income level, a higher proportion of people in any country is likely to have access to sanitation today than in the past. The same can be true of progress in reducing air pollution.

Concentrations of sulphur dioxide are lower today than in the past, so that someone living in a country with a per capita income level of $ 500 is more likely to breath cleaner air than in previous decades. The adoption of environmental policies and the investment and technological innovations induced by such policies imply that the environmental mistakes of the past do not have to be repeated. During the working lifetime of children born today, the population of the world will almost double. By the middle of the next century almost one-third of the world's population will live in countries with a population density of more than 400 per square kilometer—equivalent to the density of the Netherlands or the Republic of Korea today. The next generation will also see the size of the world economy triple.

Under simple extrapolation of current practices, this growth would lead to severe environmental degradation. Yet in virtually every economic sector, environmentally less damaging practices are available and are in use in a number of countries.

For almost every challenging water supply and sanitation, or energy and industrial output, or food production, there are possibilities for, growing more sustainably. The challenge for water supply and sanitation will be to respond to the backlog of demand while meeting the needs of growing populations. Making clean water available to everyone in the next generation will require that service be extended to an additional 3.7 billion people living in urban areas and about 1.2 billion rural inhabitants. Since only about 1.5 billion urban residents currently have access to clean water, the magnitude of the task is apparent. For sanitation the problem is even larger; the number of urban dwellers currently served is little more than 1 billion.

For a country like Nigeria, providing access to clean water for the entire population by 2030 will imply increasing the number of urban connection by four times and the number of rural connections by almost nine times. To prevent the number of people without access to adequate sanitation form rising, the population covered will have to increase to 6.5 times the current number.

The challenge for energy and industry will be to meet the projected growth in demand while controlling pollution. Total manufacturing output in developing countries will increase to about six times current levels by 2030. Average emissions of air pollutants per unit of electric power generated would have to be reduced by 90 per cent to avoid an increase in total emissions from this activity.

Emissions from heavily polluting industries—chemicals, metallurgy, paper, and building materials-will also require large reductions in discharges of air and water pollutants and in wastes produced if a worsening of industrial pollution is to be prevented. In the Philippines, for example, manufacturing output is likely to grow to nine or ten times the current level, and demand for electric power will rise even more rapidly. This means that many industries will have to reduce emissions per unit of output by between 90 and 95 per cent to avoid worsening pollution.

The technologies for achieving such reductions in pollution from energy and industry already exist in most instances. Many possibilities also exist for dramatic improvements in pollution prevention-switching to cleaner-burning fuels or recycling industrial wastewater, for example. Cleaner processes often yield productivity gains and cost reductions as well because they use materials more frugally.

The challenge for agriculture will be to meet developing countries' expected demand for food. Total world consumption of cereals will have to almost double by 2030. To protect fragile soils and natural habitats, almost all of this increase will have to be achieved by raising yields on existing cropland rather than by extending the area under cultivation.

There is little doubt that cultivated soils have the capacity to meet future increases in world agricultural demand so long as they are well managed. But intensification of production will involve the application of much higher levels of fertilizers and pesticides, as well as significant improvements in the allocation of water for agricultural use.

Doubling food production in India by 2030 can be achieved by maintaining past rates of crop yields but will require a fourfold increase in fertilizer application. By 2030 average yields in India would have to reach the level of those in China today.

Such gains in food production increase the risk of soil degradation, misuse of pesticides, spillovers from chemical applications, and excessive drawdown of water. Techniques such as integrated pest management, minimum tillage, agroforestry, integrated crop and livestock management, and soil enriching crop rotations will be needed to reduce land degradation and increase yields.

This will often require better-educated farmers, and sometimes social changes as well. When governments are committed to allocating resources to research and extension services and to providing undistorted incentives, many farmers are quick to adopt these less damaging practices. Without technologies and practices that can be applied at reasonable cost, environmental improvement is difficult. But without the backing of appropriate policies, even the most environmentally helpful technologies and practices will not necessarily be applied, unless (as is often the case in industry) they are more productive than existing methods.

The principles of sound environmental policy are well understood. But they are difficult for national governments to introduce and are even more difficult to translate into international agreements. National governments may be reluctant to challenge those who cause environmental damage; they are likely to be the rich and influential, while those who suffer most are often the poor and powerless. If institutional obstacles to addressing national environmental problems are large, they are even greater for international problems such as green house warming and the preservation of biodiversity.

It may be difficult to reach agreement among many different countries, each of which may perceive its national interest differently. If countries do not think that the benefits of agreement are worth more to them than the costs of refusing to cooperate, they may be willing to join only if other countries are willing to compensate them for doing so.

The challenges facing this generation are formidable. Many countries have not yet achieved acceptable living standards for their people. Economic growth that improves human welfare is urgently needed. Protecting the environment will be an important part of improving the wellbeing of people today, as well as the well-being of their children and grandchildren. This Report suggests a threefold strategy for meeting the challenge of sustainable development.

Build on the positive links. Policies for growth promote efficient use of resources, technology transfer, and better working markets -all of which can help in finding solutions to environmental challenges. Rising incomes can pay for investments in environmental improvement. Policies that are effective in reducing poverty will help reduce population growth and will provide the resources and knowledge to enable the poor to take a longer-term view.

Break the negative links. Rising incomes and technological advances' make sustainable development possible, but they do not guarantee it. Usually, additional incentives that capture the true value of the environment will be required to induce less-damaging behaviour. Effective environmental policies and institutions are essential.

Clarify and manage the uncertain links. Many relationships between human activity and the environment remain poorly understood, and there will always be surprises. The response should be investment in information and research and the adoption of precautionary measures, such as safe minimum standards, where uncertainties are great and there is a potential for irreversible damage or high costs in the long run. Out of the intricate web of factors influencing Third World development, environmental considerations are now assuming a priority position.

The impacts of soil erosion, salinization, deforestation, diminished genetic diversity, reliance on pesticides, and increased atmospheric carbon dioxide, among other factors, are seriously constraining prospects for development. The constraints that environmental degradation places on development are so apparent that environmental considerations, once viewed as luxuries that developing countries could ill afford, are now necessities no country can ignore.

Environmental flaws in development projects often have immediate effects, but many environmental costs take years or decades to appear. In the push to achieve economic, political, and social sustainablility, development agencies and the multilateral development banks (MDBs) have paid scant attention to long-term environmental costs.

Developing countries are littered with the rusting good intentions of projects that did not achieve social or economic success; ill-conceived projects are now building even more impressive environmental monuments to failure in the form of sediment-choked reservoirs and desertified landscapes. Success is eluding the traditional approaches to development. A recent report by the U.S. Agency for International Development (AID) noted that, all too often, foreign aid has fostered dependency rather than development towards independent prosperity.

New approaches and new ideas for international development are needed that will enable all nations-including the developed countries-to achieve sustainable development.

In 1987, the World Commission on Environment and Development defined sustainable development as that which "meets the needs and aspirations of the present without compromising the ability of future generations to meet their own needs." Successful development projects do exist, and, though the examples are few and the risks inherent in labelling any of them successful must be acknowledged, several common themes point to approaches and ideas that are capable of transforming the bleak record of international development. The Environmental Policy Institute and the National Wildlife Federation recently examined 20 case studies of development projects that appear to be achieving success in attaining sustainable patterns of resource use. The findings of that study provide insight into approaches towards sustainable development.

Environmental degradation exacts a price in any country, but the price is most apparent in developing nations, where it directly reduces the supply of food and fuel for the rural poor. Prospects for sustainable agriculture in the Third World are especially constrained by soil erosion, nutrient depletion, and salinization.

For example, 40 percent of the productive capacity of Guatemala's is land has been lost to erosion; in Haiti, the loss has been so great that no top quality soil remains. In Africa, without conservation measures, 16 percent of the rainfed cropland of 1975 will be lost to erosion by the of this century.

In developed countries, the loss of soil productivity is often remedied through fertilizer use, but intensive fertilization is prohibitively expensive for most farmers in developing countries. In India, soil erosion takes with it nutrients worth $ 6 billion (U.S.) per year in chemical fertilizer replacements. As soil productivity declines, the pressure to exploit remaining resources mounts, and further decline in productivity soon follows. Worldwide, more than half the forestland cleared annually can be attributed to replacement of degraded agricultural soil; degradation of arable land will depress food production between 15 and 30 per cent from 1975 to 2000. Other examples of development that has faltered because environmental constrains were ignored are colonization projects and livestock development programmes.

Throughout much of the tropics, forestland converted to agricultural and livestock uses loses its productivity extremely quickly (in one to three years). These environmental constraints can be ascertained during project planning; their impact during implementation of a project is seen readily in high rates of emigration from the area and low economic returns from ranching endeavours. Yet many such projects proceed at full speed in seeming' ignorance of the environmental consequences. In the Polonoroeste project in western Brazil, for example, colonization was followed by 80 per cent attrition, and the majority of the ranching endeavours were kept in operation only because of large government subsidies.

In developing countries, pesticide use rivals large dams as the most extreme example of environmental abuse through inappropriate technology brought in with development funds. With the encouragement of MDBs, high-yielding crops that require fertilizers, irrigation, and pesticides continue to be introduced in developing countries.

Although there is some dispute whether green revolution technology has produced inequitable distribution of benefits, high-yielding crops have played a significant role in the alleviation of food shortages and must continue to do so. However, the increased reliance on pesticides, often encouraged in developing countries by government subsidies, creates serious environmental and economic problems. Use of pesticides in the Third World now amounts to 10 to 25 percent of the global total and is growing rapidly. With the support of MDBs, developing countries are joining developed countries on the "pesticide treadmill," which ensures that the net gains of pesticide use will be captured not by farmers but by the pesticide industry. Between 1964 and 1974, pesticide use in Africa increased by a factor of 10; in the Philippines, it jumped fivefold just between 1972 and 1978. The cost, as measured by direct human suffering, is substantial. In Central America, chronic and acute pesticide poisonings are considered one of the region's most pervasive health problems.

The information and technologies that form the basis of ecologically sustainable development already exist. New technologies and further environmental research will aid development, but considerable success can be achieved now by incorporating already developed strategies.

To meet the needs of increasing populations, existing systems must be used more efficiently and the productivity of agriculture, forestland, rangeland, and fisheries must be improved with better management. Significant improvements in efficiency can be made in numerous areas, from industrial uses of raw materials to the harvesting and processing of foods, but the potential benefits are most obvious in the energy sector.

In 1981, for example, developing countries with low incomes spent 61 per cent of their export earnings on oil imports. Using oil more efficiently would alleviate a portion of the debilitating balance of trade deficits that plague most of the Third World. Better use of renewable energy resources would have even more direct impacts.

Biomass (mostly wood) accounts for nearly 40 percent of energy use in developing countries; increased energy efficiency in cookstoves and other apparatus could lessen the many problems associated with deforestation. Nevertheless, less than 1 per cent of all international aid has been earmarked to improve energy efficiency, and, not surprisingly, relatively few development projects take advantage of efficiency gains. The absence of energy efficiency projects from development portfolios cannot be attributed to economic or technological shortcomings. That projects stressing energy efficiency can succeed and help both the industrial sector and the rural poor has been proved with the Kenyan charcoal stove.

Energy-efficient wood and charcoal stoves relieve pressures on forests, ease the workload of rural women, and stretch the income of the urban poor. Great advances can be achieved in agricultural, forest, and range productivity in all developing countries at extremely low capital costs through soil and water conservation techniques, inter-cropping, agroforestry, and organic fertilization.

Many examples can be cited of the large, untapped potential for sustainable increases in agricultural, forest, and range productivity through inexpensive changes in agricultural practices.

In the Majjia Valley in Niger, the decline in crop yield was halted by the establishment of windbreaks that reduced soil erosion and increased soil moisture. The windbreaks provided further benefits as sources of fuelwood and construction materials. Near the village of Sukhomajri in India, a seemingly uncontrollable cycle of overgrazing and soil erosion that had left vegetation covering only 5 per cent of the ground was reversed by the construction of small dams for irrigation and a community agreement to restrict grazing. Similar successes have occurred in Kenya, where 33 per cent of erosion prone farms were terraced between 1974 and 1985, and in the Yatenga plateau of Burkina Faso, where stone lines oriented along contours to slow runoff have increased crop yields by an average of 50 percent.

In the Guinope area of Honduras, the introduction of contouring and organic fertilizers increased crop yields by up to 400 per cent. Local farmers are often the best resources for sustainable techniques suited to the local conditions. Agroforestry, for example, is not a new technology; it is often a traditional practice that farmers have abandoned because of inappropriate systems of forest or land tenure. Indeed, local knowledge may provide low-cost solutions that standard agricultural research would never discover. For example, in tropical forests, rapid soil nutrient depletion causes farmers to abandon fields after only two to three years, yet, in the southern part of the state of Para in Brazil, the Kayapo people are participating traditional methods of soil enrichment and cropping that allow them to use cleared plots for 11 years, with just a five year fallow period. In regions unsuited to the transfer of high technology agriculture, development agencies must foster a "local knowledge" revolution in lieu of a green revolution. The potential benefits of low-cost, high-gain improvements in productivity must not be underestimated, but the most difficult problems in the quest for ecologically sustainable production will be encountered as productivity is extended beyond its current limits. It is imperative at the outset to identify sustainable technologies that can enhance productivity and avoid the introduction of unsustainable ones.

Case Study: (1)

The Kenyan Jiko Charcoal Stove

In Kenya, 83 per cent of urban households and 17 per cent of rural households use charcoals stoves, known as jikos. In 1980, some 47,000 tons of charcoal were burned in household cooking, and the demand is increasing by 6.7 per cent annually. A household in Nairobi with one wage earner typically spends more than one-fifth of its cash earnings on charcoal. The traditional stove design, introduced in the early 1900s, is a metal cylinder with metal legs, triangular flaps to support a cooking pot, and a door for ash removal. In the late 1970s, several improved stoves were developed, including all metal versions and traditional jikos modified with a ceramic lining. In 1981, the government of Kenya and a local nongovernmental organization began a project promoting the dissemination of improved woodstove technology. The project's initial goal was to manufacture and sell *5,000* improved jikos by *1985.* In fact, 13,000 stoves were sold by the end of the first full year of operation, and by mid *1986* the new industry was well institutionalized and *125,000* stoves had been purchased. It is estimated that the new industry has captured 10 percent of the traditional jiko market. Nationwide, annual savings in fuel amount to 1.5 million tons of wood, worth approximately $2 million (U.S.) per year.

Case Study: (2)

Guinope Rural Development

Much of Central America is plagued by declining soil fertility caused by soil erosion and the continual monocropping of maize. By the late *1970s,* maize yields in the Guinope area of Honduras had declined substantially, largely as a result of the loss of much of the area's topsoil. Low productivity caused people to flee either to other arable land or to the capital, Tegucigalpa, and forced farmers who remained to adopt shifting cultivation. In 1981, World Neighbours, a private voluntary organization for Third World development, began an agricultural development programme in Guinope and three surrounding villages in an effort to end the cycle declining productivity with simple technologies that could stem erosion and restore land fertility. The programme introduced soil conservation practices already is use elsewhere in Central America, such as drainage ditches and contour ditches, grass barriers, and rock walls, and taught fertilization methods such as using chicken manure, intercropping leguminous plants, and spreading some chemical fertilizers. In the first years, the yields of farmers adopting the techniques tripled or quadrupled - from 400 kilograms per hectare to 1,200 to 1,600 kilograms, In the next five years, 40 other villages requested training in the soil conservation practices. This obvious success was achieved with no subsidies or donations; all costs of the agricultural production were carried by the farmers. With the spread of the programme into other villages, the benefits now cover an area stretching from the Nicaraguan border halfway to Tegucigalpa.

It is easy to see why this is the case, Most environmental modifications—whether one is altering water availability through irrigation, nutrient composition through fertilization, or pest populations through insecticides have very undesirable side effects. In *contrast,* production systems can often be enhanced safely through relative minor changes in species composition and management. For example, alley cropping, in which food crops grow between pruned hedgerows of nitrogen fixing trees ' enhances production systems with little risk to the environment. More than any other type of environmental destruction, species extinction represents the ultimate disservice to future generations. Because of human activities, the rate of species extinction is now hundreds of times higher than the background, or natural, rate of extinction. Scientists can only guess at the opportunity cost of species extinction, but no doubt it is large.

Development often has been viewed as incompatible with the preservation of genetic and species diversity, yet this need not be the case. For example, in Brazil, the value of tropical rain forest as an extractive reserve exceeds the land's value for agriculture or grazing.

More than 1 million people living in the Amazon meet their subsistence and commodity needs through small-scale agriculture, hunting, fishing, rubber tapping, and nut gathering in the natural forest. These activities are not only economically justified relative to alternative land uses, they also protect genetic resources and provide undisturbed ecosystems for the indigenous people of the Amazøn. One tool that will become increasingly important in maintaining the world's biological diversity and inbalancing development with environmental integrity is restoration ecology.

As population growth slows and as policies and institutions for managing common resources mature, opportunities to restore highly degraded ecosystems to something approaching their natural state will arise. The Guanacaste National Park in Costa Rica may become a pioneering example of the potential for restoration ecology. From the outset, project managers have attempted to-integrate restoration of this tropical dry forest with the needs of local people. The park will benefit the nearby residents through watershed protection, income from tourism, and employment. Moreover, with educational programmes directed toward school-children, civic groups, and tourists, the park will gain from the increased local awareness of the importance of conservation while the people gain a better understanding of their environment.

Similarly, an AID-funded project in Niger restores a natural forest (the Guesselbodi) after determining that the economic and social benefits of the natural forest would exceed those of plantation forestry. The land had been largely deforested before the project began, but, with water-harvesting techniques and mulching, vegetation was re-established and excessive erosion curtailed. Natural forest management at Guesselbodi will provide a 48 per cent annual return on the investment required for establishing water catchments and planting trees; this return is more than six times that of forest plantations in the region. Although these uses of natural resources successfully meet people's immediate needs, their most significant attribute might be their role in preserving biodiversity. The most important international institutions influencing Third World development are the four multilateral development banks: the World Bank, the Inter-American Development Bank, the Asian Development Bank, and the African Development Bank. Though MDB loans to governments and governmental agencies are dwarfed by those of commercial banks, the impact of MDB lending is much more significant. MDB loans require government contributions and often attract confinancing from commercial banks. In addition, MDBs sometimes offer low-interest "soft loans" when commercial financing is not available and finance economic structural adjustment programs under agreement with the International Monetary Fund.

All member countries participate in decisions regarding MDB loans, but voting power is weighted by each country's financial contribution to the bank. In 1987, for example, the U.S. share of World Bank votes amounted to 19 per cent, while Japan, France, West Germany, and the United Kingdom each contributed 5 to 6 per cent. The World Bank, source of approximately 70 percent of MDB loans, provided $ 17.6 billion in assistance to developing countries in 1986. No other MDB gave more than $3 billion, and AID, the largest bilateral aid agency, provided $6.8 billion (which included $4.9 billion in "security supporting assistance" funds, given for political purposes and only partially used for development assistance). MDB's by virtue of their size and stature, must incorporate the lessons of success if humans are ever to achieve sustainable development. MDBs have the financial and human resources to set an example for successful development policy. Yet MDBs and bilateral aid agencies are only now beginning to consider seriously the environmental impacts of development projects. The banks may have adopted sustainable development as their new slogan and embraced the importance of environmental considerations, but this change of heart is only beginning to be reflected in their lending activities. Real development in the Third World depends not on scattered successful projects but on changes in incentive structures, particularly economic incentives, that foster grassroots adoption of sustainable technologies. Development projects that succeed often do so in spite of policies that stifle sustainable development, and their success can often be attributed to innovative methods of circumventing policy and institutional constraints.

Tax policy, land reform, land titling, credit policies, commodity prices, tariffs, utility rates, social security programmes, low-income, housing policy, consumer food subsidies, and labour legislation each have at least as much effect on environmental conditions as environmental policy does. In nine developing countries in Asia, Africa, and Latin America, for instance, subsidies for pesticide use range from 15 to 90 percent of full retail cost, with a median of 44 per cent. Integrated pest management techniques will not become established while limited funds are used for subsidizing pesticide use rather than for supporting research into biological control. Similarly, attempts to slow deforestation or encourage agroforestry will not succeed while forest and tax policies promote exploitative mining of forest resources and while agricultural policies force peasant farmers to clear forested areas. In Brazil, generous tax and credit incentives have led to the conversion of nearly 9 million hectares of forest into cattle ranches; many of these ranches have since been abandoned, and most of the rest are only marginally profitable. In both the developed and developing world, commodity prices and trade policies that depress farmers' profits reduce the demand for labour and lower agricultural land prices. Consequently, the return on investments for land improvement and soil conservation is reduced: Erosion and declining soil fertility are the outcome. At the moment, the basic model used to evaluate and justify investments by MDBs and bilateral aid agencies in development projects is seriously flawed. The typical investment model does not consider the implications of irreversibility, such as the loss of arable land through salinization or desertification. Projects generally are designed for —relatively short periods of time, usually less than 10 years. The model bases the evaluation of the economic return of projects on a net present value that is calculated by assuming that the projects' future benefits are less valuable than current ones. The model also discounts future environmental and health costs. Another policy change that is overdue involves the role of women in development. National government, bilateral aid agencies, and MDBs have long ignored women's rights and their entire economic role in society, particularly in Africa, where most agriculture, fuel-wood use, and wood and water collection is done by women, this oversight is inexcusable. Women's basic rights be observed. Instead, development planners must acknowledge that women are the primary economic force in many countries, and assistance must be tailored to this reality. Successful development requires recognition by MDBs, bilateral aid agencies, and national governments that their paternalistic approach to development is often counterproductive. Planners have long ignored the importance of including the people most affected by a project—local residents—and knowledgeable nongovernmental organizations (NGOs) in the planning process. Their participation has both moral and practical implications. Planners should not assume that they know people's needs and wishes, and certainly development will not be sustained unless the needs of people are identified and unless local residents support the project at a grassroots level. Moreover, local people and concerned NGOs are often the best sources of information for the design and implementation of rural development projects. The projects most likely to be sustainable are those that have involved local residents from the outset and fostered grassroots support. For example, the Sukhumairi watershed project in India mentioned earlier succeeding because of extensive community participation in design. In the Guinope rural development project in Honduras, grassroots organizations formed as part of the extension (training) process will ensure that project achievements are not lost when funding ends.

Similarly, the U.S. Peace Crops in Zair developed an effective aquaculture programme that fostered the formation of groups of fish farmers. These groups assumed the role of the extension when the Peace Crops volunteers left. Local participation ensures that successful projects will become institutional and that the projects address people's real need. Many projects have founded because they have tried to solve problems that were critical in the minds of planners only-but not to the local people.Those who should be involved include not only the communities immediately affected by the development projects but also private voluntary organizations (PVOs), NGOs, farmers' organizations, cooperatives, schools, universities, and private entrepreneurs; all of which constitute unique sources of information useful to planning and implementation. Increased public participation brings more information to bear on the planning process. If the goal of development is to do something as quickly as possible, regardless of whether it works, then greater public involvement could be viewed as a threat. However, because the goal is successful and sustainable development, public participation is critical to solving the "problem" of development.

Case Study: (3)

Brazil's Extractive Reserves

The Brazilian Amazon contains roughly 20 per cent of the Earth's plant and animal species. Still the largest expanse of—tropical forest in the world. Amazonia and its indigenous people are under increasing pressure from land clearing for resettlement programmes and cattle ranching. Until recently, plans for the development of the Amazon largely ignored the fate of those already living there. Along with indigenous peoples, some 500,000 rubber tappers (seringueiros) live in the Amazon. These people are descendants of northeast Brazilian immigrants who colonized the forest in the late 1800s to tap latex from the wild rubber trees. The principal sources of cash income for the seringueiros are rubber, Brazil nuts, and other forest products. Each family meets its subsistence needs through small-scale agriculture, some livestock production, hunting, fishing, and gathering other forest products. This use of the forest has proved to be an extremely stable system, as many areas have been continuously worked for more than 40 years. The idea of extractive reserves was first articulated at a 1985 rubber tapper meeting, but no legal instrument existed under Brazilian law to create such a reserve. On 30 June 1987, the Brazilian Ministry of Agrarian Reform and

Development created the legislative apparatus that permits designation of state land as an "extractive settlement". Communities can be granted a long-term (up to 30 years), renewable contract for use of the land. Clear cutting is prohibited, but land can be cleared for subsistence agriculture. Extractive uses of the Amazon have proved to be sustainable and provide a model for the compatible roles of development and conservation. Reserves provide direct economic benefits to the seringueiros, undisturbed land for indigenous people, and invaluable protection for genetic resources. Over a 10-year period, the net present value of extractive reserves seems to be slightly lower than that of either cattle ranching or agriculture. However, because neither ranching nor agriculture can be sustained in the Amazon, the net present value over 1.5 to 20 years favours extractive use of the forest. It is the nature of large institutions to be inflexible. It is the nature of successful development projects to respond in an adaptive, experimental fashion to local conditions and constraints. Herein lies one of the more pervasive controversies in development.

MDBs and bilateral aid agencies support projects on a scale that generally prevents adaptive response to local conditions. PVOs undertake small projects that can be modified to match specific community needs. Several PVO projects, especially recent agroforestry efforts, have been strikingly successful whereas MDB and bilateral aid projects, which potential could improve the well-being of a significant number 0 people, have been strikingly unsuccessful. The solution is not to channel all aid money through PVOs, although it is clear that a much greater percentage should go through them and that requirements for matching funds should be eased. Many aspects of development including the need for basic policy reforms, industrial sector development, and housing programmes—are beyond the scope of PVOs. Moreover, PVOs may have no advantages over MDBs and bilateral aid agencies in managing very large rural development projects that only indirectly involve the rural poor (such as large farm plantations of livestock ranching). However, in agroforestry, for example, where community-based participation is a prerequisite for success, microscale administration of projects is critical.

Case Study: (4)

Agroforestry Outreach in Haiti

Haiti with a per capita gross national product of $377 (U.S.) and an average rural annual income of less than $100, is the poorest nation in the Western Hemisphere. Many of Haiti's environmental problems can be traced to deforestation. In 1950, four fifths of Haiti was forested. Today, forests cover less than 8 per cent of the country. In 1981, the U.S. Agency for International development (AID) funded an $8 million Agroforestry Outreach Project (AOP) to promote trees as a cash crop. The project design grew directly from lessons provided by the few successes and many failure of 19 previous erosion control projects in Haiti. AOP initiated a tree-growing programme that was to be profitable for the peasants. Funds for the project were distributed to three private groups, one of which acted as an umbrella organization for more than 170 smaller Haitian and U.S. nongovernmental organizations (NGOs). Through the use of this intermediary NGO, AID was able to avoid the added administrative costs of dealing with many small I NGOs, while the small organizations avoided the bureaucratic requirements of loan and grant recipients. By 1986, 39 nurseries were in operation, producing more than 5 million seedlings per year. Approximately 110,000 farmers had planted more than 25 million seedlings with a survival rate of 50 per cent. Tree planting has diversified farm production, increased farmers incomes through the Sale of wood and tree products, buffered them from droughts that could decimate annual crops, and allowed the sustainable use of marginal land. Soil erosion has been reduced and the fertility and mφisture of the remaining soil has increased. One most promising model is provided by the Haiti Agroforestry Outreach Project. Through the use of an intermediary NGO, AID has avoided the administrative costs of dealing with 170 NGOs, which in turn have avoided the agency's bureaucratic requirements. The potential impact of NGOs is, large: A Club of Rome study found that in Latin America some 25 million peasants have benefited from NGO projects; in Africa, 12 million; and in Asia 60 million. MDBs and bilateral agencies must also become more flexible in project implementation.

Development debacles are often apparent quite early in the project cycle, but the financial and institutional inertia of most large projects obstructs experiential learning and modification. This need not be the case. The Haiti Agroforestry Outreach Project funded field studies of farming communities during the design stage and included a two-year applied research component during project implementation. This learning process can be helped by including NGOs and local residents in the planning process from the outset and by providing rewards on the basis of project success rather than on the amount of loans granted or the timeliness of project completion. Project design should be treated as a continual activity during implementation. National environmental assessments, already being conducted by the World Bank and bilateral aid agencies, should receive greater attention and be more thoroughly incorporated into project planning. Pilot projects and studies to evaluate the potential effectiveness of new technologies and extension methodologies should receive much greater emphasis. In projects that include aggroforestry components, for instance, it may take three or four years activities such as species testing, nursery establishment, and training programmes for extension staff to develop before they can support extensive activities.

A five-year project cycle is not suitable for this type of project, and more consideration must be given to phased implementation following a pilot project. The very best technologies and management systems are useless unless they can be disseminated to those who will benefit from them.

Successful extension methodologies vary widely. The Guinoperural development project worked well by using local Honduran farmers who provided hands-on training in others' fields.

In contrast, a successful organic farming project for small farms in Chile brought farmers to central demonstration farms for training. Highly technical methodologies may need outside experts as extension agents for as long as four to eight years, whereas the organic farming and Guinope projects rapidly trained farmers as extension agents.

Notwithstanding these varying roads in success, two generalizations can be made. First, adherence to one kind of extension will restrict the type and range of projects that will succeed. Some PVOs achieve repeated success in part because they select systems to which their extension techniques are well suited.

Many MDBs projects fail because they attempt to use one extension methodology in all situations. Second, although subsidies always increase the rate at which new technologies are adopted, the benefits of rapid dissemination seldom extend the costs of the economic distortion caused by the subsidy.

Technology make an essential contribution to the material and social welfare of modern society. Indeed, technology is an integral part of human development. Humans may even be described as "technological animals" differing from other tool-making animals in their more systematic use of technology.

Technology may often be regarded as a recent phenomenon, as are environmental destruction and occupational injuries. However, this concept is false mainly because the history of technology is almost as long as that of human history.

The history of technology is also the history of its side-effects, covering the environmental and occupational science. In ancient history, information has been found on diseases caused by primitive technologies as in the works of the classical scientists Hippocrates, Plinius and Celsius, and poets such as Martialis and Juveniles, who reported the dangers of work with metals and sulphur. Plinius especially seems to have been the first to describe the use of personal protection devices, describing the use of face masks made of bladder skin to protect workers against inhalation of dust.

Ancient technology could be characterized as small-scale technology, since the associated occupational hazards were fewer in number and intensity than in later centuries.

With an increasing demand for raw materials, technology developed rapidly during medieval times, a trend reflected in the appearance of new occupational diseases. In particular, mining and metal technologies had been a dangerous industry for a long time. A well-known case may be women living in the Carpathian mine districts, who remarried several times because their husbands became victims of "metal diseases".

During the 18th century, the agrarian and feudal Western societies were changed by the industrial revolution. This process may have reduced some occupational hazards; however, it also introduced a number of new ones.

Developing modern technology has, during the last 200 years, enlarged the spectrum of hazards following the onset of new industrial activities, such as iron and steel smelting and forming production, pulp and paper factories, and oil extraction refineries.

A rough global estimate of current occupational accidents and diseases suggests that tens of millions of workers are injured every year of which probably more than 100,000 are hilled. It has also been estimated that 5 percent of the total working days of a nation could be lost because of occupational injuries.

However, international comparisons of occupational hazards are hard to make, because information from different countries is not based on internationally-agreed criteria. Technological development is a dynamic force changing the patterns of occupational hazards, resulting in a complex picture of the risk panorama.

A positive feature is that the frequency of accidents has decreased in many countries, particularly in mining, construction and transportation industries, while accidents seem to be increasing in agricultural and forestry work. As advances are made in the scientific study of the workplace, new occupational hazards such as noise, vibration, chemical hazards, and physical and mental strain, have become increasingly evident problems. Noise, dust and chemical air pollutants are the factors mainly responsible for the increase in the number of disabilities caused by occupational diseases.

INDEX